BEEKEEPING IN VICTORIAN NOTTINGHAMSHIRE
1837 - 1901

John Stuart Ching

Northern Bee Books

BEEKEEPING IN VICTORIAN NOTTINGHAMSHIRE

© John Stuart Ching

All rights reserved. No part of this publication may be reproduced, stored in a retrieval system, transmitted in any form or by any means electronic, mechanical, including photocopying, recording or otherwise without prior consent of the copyright holders.

ISBN 978-1-912271-23-8

Published by Northern Bee Books, 2018
Scout Bottom Farm
Mytholmroyd
Hebden Bridge HX7 5JS (UK)

Design and artwork by D&P Design and Print

Printed by Lightning Source UK

BEEKEEPING IN VICTORIAN NOTTINGHAMSHIRE

1837 - 1901

John Stuart Ching
BA, BA (Hons), PGCE, MEd
Archivist, Nottinghamshire BKA

Queen Victoria and Prince Albert inspect the beekeeping exhibit at the Great Exhibition of 1851

FOREWORD

I am very pleased to have the honour of writing a Foreword to this important work on Victorian beekeeping in Nottinghamshire by Stuart Ching, exceptionally so, as I have never been a member of Nottinghamshire Beekeepers Association. However, for over a quarter of a century, I lived just across the Trent from Nottinghamshire in Lincolnshire, and no doubt with favourable winds and forage, my bees would have flown over the boundary for their vital stores. Not only that though, I had many beekeeping friends in Nottinghamshire and regularly attended the annual auctions of bees and equipment of the NBKA, as well as the Nottingham Show, both events being at the Show Ground in Newark. It was in Newark, too, that I had a honey stall for over eighteen months in the old Butter Market each Saturday and Public Holiday, and there met many people who had a great interest in bees and honey.

Stuart Ching's fascinating book is the result of diligent research into the contemporary accounts of beekeeping in the county during the long years of Queen Victoria's rule, an important time both socially and in terms of the development of the craft of beekeeping. For a long time during the Queen's reign there was little in the way of education for the masses until the setting up of Board Schools in 1870 which was meant to provide education for all children between the ages of 5 and 13 (fees had to be paid by parents, though the Board paid for the children of poor families). However, perhaps not surprisingly, there was much opposition to this initiative as the Church, which in many circumstances provided education for the poor (funded by the State), did not wish to relinquish neither the revenue which they received nor their influence over the children. Also, of course, there was the fear that the educated masses, those of labouring backgrounds would, once enlightened, become dissatisfied with their lot which could lead to social unrest. However, despite these concerns, education until the age of ten became compulsory under Gladstone in 1880, and became free for all pupils in 1891. This state of education existed until the year after Queen Victoria's death in 1901, when the Balfour Act of 1902 put the provision of education into the hands of the local authorities.

It is important to understand this in the context of Victorian beekeeping; how the new knowledge and systems of management were passed down to those beekeepers with little education (who no doubt, though, had good practical skills).

Undoubtedly, the setting up of local beekeeping associations was fundamental for the dissemination of information to beekeepers. Whilst Nottinghamshire Beekeepers Association was initially started in 1879, and then permanently set up in 1884, it is likely that many beekeepers in the labouring classes would have had little in the way of formal education so thus benefited by their attendance at meetings.

Reading through the earlier pages of Stuart's book, it is very noticeable that three distinct classes of beekeepers are referred to: the landed gentry type; the intelligentsia - particularly the clergy; and the labourers. Each class seems to have its own defined place in the beekeeping hierarchy, with the clergy notably dominant in their roles as chairmen or secretaries of the associations. Indeed, as regards the labouring people, the title of 'cottagers' is frequently bestowed upon them - with special classes being open to them in honey shows, for instance, as well as the type of hives in which they eventually kept their bees.

I find it extremely interesting that one of Stuart's main sources is the British Bee Journal, which was first published in 1873 - that is, eight years before George Newnes, the father of popular journalism, launched his very successful Tit Bits, which eventually led to the publication of the Daily Mail and the Daily Express: clearly those who had benefited from the Education Acts had a great thirst for knowledge.

In the wide-ranging contents of the book itself, the many longer pieces are complimented by other shorter items which add greatly to this marvellous picture of Victorian beekeeping. For example, the theft of hives was rampant during those days and those responsible were severely dealt with by the courts. Certainly jail terms were given to offenders - but with hard labour too and with part of the time being spent in solitary confinement. Would such punishments, I wonder, deter the bee rustlers of today?

Stuart Ching's book is not only of interest to Nottinghamshire beekeepers, but to all those who have an interest in the history of beekeeping. So many fascinating items which would be difficult to find are present in this excellently researched book. Let's hope that his work inspires other associations that have not yet written up their histories to do the same.

John Phipps
April 2018

Stuart Ching's series "The Great War – Its Effect on Beekeeping - as seen through the pages of the British Bee Journal" has been published in The Beekeepers Quarterly, since 2014.

BEEKEEPING IN VICTORIAN NOTTINGHAMSHIRE

INTRODUCTION

The Nottinghamshire Beekeepers' Association (NBKA) was founded in 1879 but soon fell into disarray. The second attempt at forming an Association in 1884 was more successful and it is that organisation which continues to this day.

NBKA has a poor reputation with regard to its records. Few original documents exist from its early start. Indeed, there is a story that one of its first Minute Books was later found for sale in a second-hand bookshop in Stamford, Lincolnshire! Where I have been able to obtain information from Minute Books it is noted in the text. However, as this information was hand-written in ink sometimes this was quite difficult to read. Unfortunately the correspondence noted in the Minute Books has not survived.

The documentary history of NBKA did not really start until the production if it's first newsletter in 1958 but, apart from the first issue of BEE-MASTER as it was called, no complete issues survived until 1968.

I have access to the British Bee Journal (BBJ) from 1873 – 1922. This publication by the British Beekeepers' Association (BBKA) covers the time from the original ill-founded attempt to form an Association but did not completely cover the gap up to the date of the first newsletter or, indeed, the dates of the subsequent newsletters. I have also been able to search, through the British Newspaper Archives, the local newspapers of the relevant times. The Nottinghamshire Evening Post (NEP) and the Nottinghamshire Guardian (NG) are good sources of beekeeping activities in the county as they often gave differing reports of the same event. The NG ran an anonymous "Beekeepers' Corner" for some time.

BEEKEEPING IN VICTORIAN NOTTINGHAMSHIRE

RESEARCH

This history is written in chronological order as extracts will be taken from sources which are in themselves arranged in such an order (dates of publication). As people are named in the extracts a diversion will be made to include biographies of them. This will not be possible in all cases but those who made a contribution to the well-being of NBKA are worthy of some acknowledgement.

Access to the various census records via the Internet has enabled me to find dates for some of the people involved with NBKA. Unfortunately this has not been achieved with all of them which I regret. This task is not helped by the capricious nature of the spelling of names in many cases. It is easier to find the "rich and famous" (mainly nobility, politicians or clergymen) for inclusion in a work such as this but an effort was made to find as many of the "ordinary" members who have supported the Association over the years. It is hoped that, by publishing this book, more such members may be identified.

No attempts will be made to equate historical currency values with modern ones.

Perhaps it is appropriate here to explain about beehives. Until about the mid-1880s the term 'hive' or 'beehive' meant straw skeps. Wooden hives were being introduced "on the Nutt's principle". These had moveable frames on which the bees built their comb. As these were (and still are) heavy where the report says "stole three hives" these must have been straw skeps as no one person could carry a wooden hive full of bees and honey more than a few metres.

Once the Association had been formed and honey shows began to be held it will be obvious from the show schedules that, at first, skeps and wooden hives could both be entered. Skeps were soon eliminated from shows as the Association had as one of its aims, "the reduction of the cruel practice of using sulphur to kill the bees." This process required skeps to be held over a pit of burning sulphur so that the bees were killed and the honey recoverable. What the honey smelt and tasted like is not recorded! The BBKA, at one time, held its meetings in the headquarters of the Association for the Reduction of Cruelty to Animals (which became the RSPCA)

ACKNOWLEDGEMENTS

During the preparation of this book many friends and members of NBKA have rallied round to give me encouragement particularly during my spells in hospital having treatment which put back it's publication for two years.

Penny Forsyth, who took over the editorship of **BEEMASTER**, the newsletter of NBKA, when I had to relinquish the post after twenty plus years, never failed to give my efforts the "thumbs up!"

The Council of the Nottinghamshire Beekeepers' Association generously agreed to cover any costs involved with the production of this publication.

I am also grateful for the support given by Jeremy Burbidge, the publisher of this work. He it was who repeatedly told me "this work has to be written for the sake of NBKA and the beekeeping world!"

It is my earnest wish that additional material will be added to this initial essay in years to come.

"Nottingham Royal" beehive made by
C Redshaw of South Wigston, Leicestershire

BEEKEEPING IN VICTORIAN NOTTINGHAMSHIRE

On 20 June 1837, William IV died in his sleep after a reign of seven years. His niece, the 18-year old Princess Victoria, inherited the throne. Her accession marked the dawn of a new era in Britain's history, which would come to represent industrial growth, scientific advances and vast imperial expansion.

Nottingham Review 21 Jul 1837
A swarm of bees settled in Nottingham Market-place last Friday afternoon, on Mr Deverall's sign, bottom of Sheep-lane. No difficulty was experienced in hiving them, and they are now in the possession of Mr D. Most probably they had migrated from some hive in the Fish-pond Gardens.

Nottingham Review 8 Sep 1837
Last Sunday night, or early on Monday morning, some thieves entered the garden of Mr George Rogers, of Arnold, and carried off a fine hive of bees. They made use of brimstone to smother the bees, and seemed quite used to the trade.

ARRIVAL OF THE QUEEN'S PROCESSION AT WESTMINSTER ABBEY FOR THE CORONATION

The Coronation of Queen Victoria took place on 28th June 1838. The procession to and from the ceremony at Westminster Abbey was witnessed by unprecendently huge crowds as the new railways made it easier for an estimated 400,000 people to travel to the capital for the event.

Not so in Nottingham where the mainline railway to London did not reach the city until 1894 - near the end of Victoria's reign. To get to the capital travellers had to go to either Derby or Newark or even Grantham as the city fathers were of the opinion that "travelling by rail would not catch on!"

Nottingham Review 30 Nov 1838
Beeston Library. On Wednesday evening, the 21st inst. Mr John Pearson delivered a very pleasing and useful address on the economy and management of bees, in the course of which the various construction of foreign and domestic hives, and the most effectual methods of obtaining honey without the destruction of this interesting insect, were well discussed and illustrated.

Nottingham Review 22 Oct 1841
Stealing Bee-hives, John Taylor, aged 31, a framework-knitter, of Hucknall Torkard, was charged with stealing, on the 18th of September, two bee-hives and thirty pounds weight of honey, the properly of Wm. Coope. Mr. Wildman called the prosecutor, who said that on Saturday, the 18th of September, he saw the bee-hives safe in his garden; on Sunday morning he was awoke by a noise, got up, and found the prisoner and Edward Hill and Isaac Allen, two county policemen, in his garden; the prisoner was in custody, but another man jumped over the hedge, and ran away. The covering of the hive was removed, and the bees were all stifled; there was a piece of brimstone burning near the hive. Isaac Allen confirmed this evidence. The prisoner, in defence, said that a man, named Geo. Holland stifled the bees, and he had nothing to with it. The prisoner had a good character given him. Verdict: Guilty. To be imprisoned six calendar months; one week in each month solitary.

Nottingham Review 19 Jun 1840
Monday, a swarm of bees took up their abode in the gas lamp at the entrance into St. James'-street from the Market-place. Several hundred persons were attracted to the spot by the novelty of a swarm of bees in a gas-lamp. A baker hastened home for a hive, which he sweetened with honey, to make it more palatable to the queen bee and her subjects; the lamp-post was then shook, and the majority of the latter dropped from the top of the lamp on to the top of the iron pillar, and there they stuck, for such was their attachment and loyalty, to their beloved queen, that they could not be induced to enter the hive; at length the queen - the object of their adoration - was hived; that very instant her subjects became willing captives, and the baker carried off his new colony in triumph. We have heard the bees came from Hyson Green.

Nottingham Review 3 Sep 1841
On Thursday week, as Mr Thomas Duke, of the Bricklayers Arms, Lombard-street, Newark, who had been brewing, was about to put his brewhouse in order, he discovered to his surprise a fine swarm of bees in the centre of the mashtub; but Mrs D. accidently dislodged them, or they would have been housed without the least difficulty.

BEEKEEPING IN VICTORIAN NOTTINGHAMSHIRE

Nottingham Review 25 Nov 1842
Two bee-hives were stolen from the garden in the occupation of Mrs Sarah Moss, Greasley, in this county, during the night of Monday, the 14th inst. Denis Bush was commited to the next Sessions for stealing said honey.

Nottingham Review 24 Oct 1844
Thursday night last, Mr. Freesby, of Belvedere-street, Mansfield, had about fifty pounds of honey stolen from his garden, in Bull's Head-lane. One hive was emptied, supposed to contain about thirty pounds. From the adroit manner in which the thieves had performed their task, it is conjectured that they are no strangers to that line of business. Another hive, that lay contiguous to the three that were emptied of their honey, was also shifted from its place, and attempted to be robbed, but the depredators were most probably disconcerted at the time by some persons passing in the lane.

About a fortnight ago, Mr. Woodward, of Rushley Farm near Mansfield, had eight valuable hives of bees stolen from his premises, which were found by the police secreted in a hovel, within a stones throw of the premises. The honey was made such work with that it would be useless, except as winter food for the bees.

Nottingham Review 15 Nov 1844
On Wednesday night, some persons entered the garden of Mr. Wm. Wright, carrier, and stole therefrom two beehives. On the same night, they also entered the garden of Mr. Samuel Greaves, cordwainer, of Calverton, and stole two beehives. In an empty house, situated in George's-lane, leading to Arnold from Calverton, the thieves made their place of resort, where they made a fire, and obtaining a puncheon from Mr. Monk's premises, forthwith proceeded to melt the honey out of the hives, but Mr. W Marriott, at the Spring Gardens, happening to come down the lane, met one of the rascals with a hive in his hand. Suspicion was immediately created that all was not right. Mr. Marriott instantly gave the alarm, and the men decamped, leaving behind them part of the hives uninjured and part destroyed. They also left a good coat, containing fifteen shillings and some keys.

Nottingham Review 15 Nov 1844
We are informed that Mr Moult, who resides at a short distance from Mansfield, near to the Derby-road, has this year taken three hives of honey weighing respectively 114 lbs, 105 lbs and 94 lbs.

Nottingham Review 9 May 1845
On the 1st the instant, the inhabitants of Worksop and Mansfield turned out of their houses to witness a cavalcade of six coaches full happy faces proceeding to and from the classic grounds of the far-famed abbey of Newstead. They were the pupils

of the Pestialozian academy at Worksop. A large swarm of bees settled upon the boys as they were about start, which was considered a happy omen as, in the language of Plautus, "Übi me,libi apes."

Relating to Johann Heinrich Pestalozzi (1746-1827), a Swiss teacher, who developed a system of elementary education incorporating manual training.

Ubi mel, libi apes – where honey, there bees. If you want support, you must offer something in return.

Nottingham Review 25 Dec 1846
A few nights back, a beehive was stolen from a garden at Hill Top, Eastwood, belonging to a man called Bradley.

Nottingham Review 26 Feb 1847
About seven years ago, a swarm of bees made its appearance at Barnby Moor, near Retford, and took its abode in the roof of Mrs. George Clarke's house. Since then other swarms have come from nobody knows where, and have joined the orginal stock. Having become so formidable and at times so irritable, as to become a nuisance, it was determined to exterminate them. Accordingly this was done on Friday, by the gardener, who was well equipped for the job. Having driven out some and killed the remainder, he succeeded in taking their honey, which weighted 3st. 11 lbs and also a proportionate quantity of wax, which he secured as the spoil of his somewhat hazardous undertaking, considering the difficult position they had selected as the place of their abode.

Nottingham Review 6 Aug 1847
Rare Collection of Bees. Mr. G. Whitworth, of Carlton, near Worksop, on the 26th of June last took 45 lbs of clear honey from one box, and on the 6th inst. 48 lbs more from a second box, a co-lateral hive, on Nutt's principle; and it is supposed by competent judges, that there are 50 lbs of honey in box and glass still remaining in the same hive. Mr. W. is well known to be a skilful man adept in the management of bees, and we believe, stands unrivalled throughout the county of Nottingham, for his extensive and valuable collection of hives, which are of a very unique description, many of them having been designed and contrived by himself. There are a great number of spiral hives (skeps?), judiciously arranged, in his extensive garden. It is really pleasing and interesting sight to see the bees most industriously engaged in prosecuting their wonted task. All the spiral and collectoral hives are furnished with bell glasses; and of course the bees are at any time to be seen busily employed.

Nottingham Review 27 Aug 1847
Swarm of Bees in the Ceiling of a House. Last summer, a swarm of bees, belonging

to Mr. Wm. Osbourn, Hucknall Torkard, hosier, left the hive, and after a circuitous route, ultimately took their abode in the ceiling of the house of Mr. William Thompson, boot and shoemaker. They obtained entrance through a crevice in the wall at the end of a beam. In this elevated situation they have lodged nearly twelve months. About a week ago, their privacy was invaded, and after no trouble, the citadal was taken, and the spoil, which consisted of a very large quantity of wax, about forty pounds of honey, and two pounds of wax, was claimed by Mr. Thompson as his own, the value of which was about £2.

NG 15 Mar 1849
The case of Finn v. Galland in March gave an incidental insight into beekeeping. The said plaintiff was a poor man owning a cottage and garden at Sutton-on-Trent, and had a right of road across a field which the Rev. Mr. Thomas Galland of Laneham had recently purchased, and had refused to allow the right any longer. He described Mr. Galland as a man who was fonder of going to law than doing a good action towards his neighbours, or he would not have put his client to the expense of coming to that court, for the costs of that trial would be more than all his property was worth.

Mr Wilmot, horse dealer, of Lincoln, who deposed that his father once occupied the house now occupied by Mr. Galland. He went there about the year 1823, and when he first went his father objected to Trussell taking coals, &c. up the road. For about a year and a half they were unfriendly, and about that time his father bought a lot of bees, and as they were brought home they got upset, and he had to send for old Mrs. Trussell, who kept bees, to put them to rights. After some deliberation, the jury returned a verdict for the plaintiff.

Probate will of Rev. Thos. Galland of Laneham, clerk of the Wesleyan persuasion
To Rev. Geo. Rigg of Lincoln, Thos. Colton, sen., of Cottingham, farmer and Wm. Hutchinson Lazonby of Manchester, Lancs all real and personal estate (except furniture for a small house for wife Sarah and d. Mary) on trust to pay:-£300 or whatever amount due to Mary Lazonby of Rampton (when trust of Wm. H.L. shall cease), and residue for wife for life, then to d.; if d. die before 21, then to children of sis. Ann Stevenson of Barnard Castle Durh. G.R. and T.C. executors. Memo. of York probate, 17 Nov. 1843.

NG 12 Jul 1849
On Monday morning last, a very fine swarm of bees settled on the eve of Mr Freeman's house in Millstone-lane, Nottingham and were safely taken by Mr Fox, bricklayer.

Millstone Lane was in the Glasshouse-street area and does not exist today.

NG 26 Jul 1849
From three stocks of bees, commonly called hives, standing in the garden of

Mr Durham, at the Manor Cottage, Worksop, there have been, since the 15th of May 1849, ten swarms of fine, strong and healthy bees! An occurrence, perhaps, unparalleled in apiarian history

NG 6 Sep 1849
On Monday evening, a person named Coggin, at Skegby, having found what he conceived to be a rabbit's nest, cautiously looked round to see the coast clear, and then plunged his hand into the hole; but to his utter astonishment he pulled out a great quantity of honeycomb and bees wrapped in the down. He was not long in decamping from the store, but not before several of the disturbed had stung him very severely.

Nottingham Review 2 Nov 1849
At Bulwell, a few nights ago, some evil-disposed person entered the garden of Mr. Thomas Chamberlain and stole two excellent bee-hives. They carried the spoil into an adjoining field, and there with matches, tried to stifle the bees. It is conjectured that they were put off, as the next morning, a quantity of matches and a thick truncheon were found in the field. Then they proceeded to Hemshall-lane, where they cut the hives into pieces with a sharp instrument. Here, we believe, they were again put off, as part of the hives and combs were found the next morning.

NG 8 Nov 1849
Samuel Voce, lace maker, born in 1810 in Cotgrave, is well known for his knowledge of the natural history of the bee, has a young apple tree growing in his yard, near to his cottage door which bears a remarkably fine fruit; and this year the fruit being finer than usual, he thought he would make a present of a peck of these beautiful apples to her Majesty the Queen. The present was kindly and graciously received, both by her Majesty and Prince Albert; and a letter was addressed by the Lord Steward to the Archdeacon Browne, rector of Cotgrave, to know the character and circumstances of the said Samuel Voce, and a satisfactory answer being returned, stating that if not in the lowest, he was in a humble station of life, and with a large family her gracious Majesty desired the Lord Steward to remit a Post-office order for two sovereigns, as a remuneration.

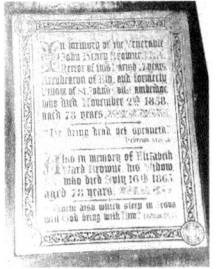

The Venerable Archdeacon John Henry Browne, was rector of Cotgrave 1811-1858 although he was also Archdeacon of Ely. This memorial plaque to him and his wife can be found in Cotgrave church.

On Scrimshire Lane, near Cotgrave Church, can be found an old wall, dubbed the "Thousand-year wall". It is riddled with small holes made by, and providing homes for, a large group of solitary bees.

BEEKEEPING IN VICTORIAN NOTTINGHAMSHIRE

NG 24 Jan 1850
On Tuesday night last some person entered the garden of the Rev. J. Thompson, vicar of Kirton, and upset all his beehives. The same night some person broke into the cellar of Mr Wm. Roberts, innkeeper, and took a quantity of ale, two lbs of butter, and three loaves of bread. The same night some person cut the gig apron belonging to Mr W Taylor, blacksmith, of Boughton.

Nottingham Review 25 July 1850
On Thursday night, the 18th inst., or early the following morning, a number of gardens in the Park were visited by marauders who obtained a large booty of wearing apparel, etc. We understand that Mr Wynne, jun., who was sleeping in the summer-house at the time, expecting a swarm of bees, was considerably discomforted, on awaking, by the discovery that the impudent rascals had decamped with the clothing of which he had divested himself on retiring to rest.

NG 4 Dec 1851
Sold By Auction on the premises of late James Nixon, Beeston, Bee Hives, with the live contents

NG 12 Feb 1852
Auction of goods of John Marriott of Lambley – set of box bee hives

NG 11 Mar 1852
Auction of goods of Welles Charlton of Chilwell – 11 hives of bees

Marble Plaque south aisle, south side in St Mary's, Attenborough, to Orton Welles 1820 (erected by Welles Charlton)

Nottingham Review 3 Sep 1852
Southwell Petty Sessions. Aug. 31. Before the Rev. JD Becher. Samuel Dennis and Henry Bartles, labourers, and George Pettner, cordwainer, all of Farnsfield, apprehended on Sunday, the 29th ult. by Pc. Fox, and Inspector Wimant, charged with stealing in the night time, a quantity of honey from a bee-hive, situate in the garden of Mr. George Bursey, of Greaves'-lane, in the parish of Edingley, farmer, and also a quantity of apples from his orchard. They were each convicted for stealing the apples; Dennis in the penalty of £1, and costs, 7s 6d and in default two months imprisonment; and Bartles and Pettner, each in the penalty of 15s and costs, 7s 6d and in default, one months imprisonment. Pc. Fox was on duty at the time, and great praise is due to him for his exertions in bringing them to justice.

NG 3 Feb 1853
Auction of goods of Mr Faulks (who is leaving the neighbourhood of Langar) - 8 stocks of bees (very good)

NG 31 Mar 1853
Sale by private contract – 9 good stocks of honey bees in straw hives and 8 in hives on Nutt's principle together with the house in which they are placed. Mr Faulks of Langar, will shew them on application.

A classic work in bee culture, Thomas Nutt's "Humanity to honey-bees" (1835) addressed both bee management and the natural history of bees. He was draper and grocer of Moulton Chapel in Lincolnshire. Writing during a period in which beekeepers were kept busy outdoing one another in innovations to improve the profitability of what was literally a cottage industry, Nutt developed one solution to the problem of extracting honey from a hive without destroying the colony. As it proclaimed in its title, Nutt's "plan" had the double benefit of greater humanity toward bees and greater profitability for beekeepers.

Presumably the sale was of a beehouse not the dwelling place of the beekeeper!

NG 15 Dec 1853
Some evil disposed person or persons entered the garden of Mr W Gilbert, farmer of Walesby, on Wednesday night last, and took a hive of bees. A few more stood by but they contrived to take the best. A sharp lookout is kept for these depredators.

NG 6 Jul 1854
Watnall Hall Auction of goods – four hives of bees.

NG 8 Feb 1855
Weather. The honeybee was seen flying on the 7th January.

NG 15 Mar 1855
A few nights ago two hives of bees were stolen from the garden of Mr Wm Elvidge in Caunton. No clue has hitherto been obtained of the thieves.

Beehive – cheap calicoes. JW Worrall, Proprietor, 25, Hockley, Nottingham Wright's Directory

Rev. JG Digges who was a member of the examining board of the Irish Beekeepers' Association and editor of the Irish Bee Journal writes: 'Sheets and quilts are required upon the frames to preserve heat; to prevent draught; and to keep the bees from ascending into the roof. The sheet is made of bed ticking or unbleached calico.'

BEEKEEPING IN VICTORIAN NOTTINGHAMSHIRE

NG 6 Sep 1855
Mr. S. Pearson, grocer, of Epperstone, hived a swarm of bees on the 5th or 6th of June last, and took from the swarm on the 17th of August 43 lbs of clear honey, exclusive of wax and refuse.

NG 28 Feb 1856
Auction of goods of the late Mr Manley, Sherwood – Three capital beehives and bees. Two bee-hives.

NG 29 May 1856
An aged widow of the name of Crampton, mother to John Crampton, shepherd, and W. Crampton, bee-hive maker, of Cotgrave, died, and was buried at Bingham a week or two ago, and left the following progeny:— 7 children, 60 grandchildren, 62 great grandchildren, and 3 great, great grandchildren, in all 132.

NG 26 Jun 1856
On Wednesday week, about four o'clock in the afternoon, a fire occurred upon the premises of Mr. Jackson, druggist, Market Place, Mansfield, under the following circumstances. Mr. Jackson was preparing a composition for making furniture cream, which consisted of a compound of turpentine, bees' wax, and other inflammable materials, and was boiling the same over the fire in a pan when it suddenly became a blaze from the vapour arising and coming in contact with the flame from the fire. Mr. Jackson immediately lifted the pan from off the fire, when he was completely enveloped in flames, which communicated with some clothes which were hanging on a horse to dry. These and other articles were speedily demolished. Owing to the prompt assistance of the neighbours and a good supply of water at hand, it was subdued without doing much damage. Mr. Jackson was severely burnt about the arms, and is under the care of M. Furniss, Esq., and progressing favourably.

NG 26 Jun 1856
A Wonderful Swarm of Bees. On Saturday afternoon last, about three o'clock, a swarm of bees, attracted by the chipping of several bricklayers' trowels whilst at work at the St. Ann's Wells, near Nottingham, alighted on a large elm tree about four stories high, when Mr Wm. Fox, of Glasshouse-street, in this town, sent immediately to Nottingham for a hive, and at once ascended the tree and succeeded in getting them safely hived, and carried them on his head to the garden, whilst they were running all over his head and face, and singular to say not one of them stung him, where they are all doing well. There are some thousands of them, and are decidedly a beautiful swarm.

NG 27 Mar 1857
Auction. Sales of goods of Mr G Wragge of Toton. Four hives of bees.

NG 1 Mar 1860
Auction. Mr George Wigley, Horse-in-Dale Farm, Cinderhill - three hives of bees.

Felony at Sneinton. John Green was placed at the bar on the charge of stealing a beehive, at Sneinton, on the 8th instant, the property of Mr. John Morley. Prosecutor deposed to missing the hive from his garden, and on the 5th instant he went into the pig-sty of defendant were he saw his hive. Sergeant Hulse took him into custody, when he said that he had bought the hive for 10s. Prisoner had nothing to say in defence. There was a second charge against him of a similar nature, and he was ordered to be imprisoned for four months with hard labour.

NEP Dec 1860
"The winter of 1860 was, perhaps, the coldest of which we have any record. On Christmas morning of that year the temperature at 4 ft. above the ground was 8 deg. below zero, on the grass, 13.8 deg. below zero, or 45.8 deg. of frost. In the Nottingham Journal of December, 1860, Mr. EJ Lowe, of Highfield House Observatory, Beeston, near Nottingham, wrote as follows :—' I herewith send you a report of, perhaps, the most extraordinary cold ever known in England.'

NG 15 Apr 1862
The members of the Midland Scientific Association met in the Museum Library, at Derby, on the 2nd inst. The president, Sir Oswald Mosley, Bart., having been prevented by his engagements from being present, the Honourable and Rev. OWW Forrester took the chair. Mr. Edwin Brown read a paper "On the plan on which bees and wasps construct their cells." This paper closed the proceedings of the evening.

Reverend Orlando Watkin Weld Weld-Forester, MA, 4th Baron Forester (1813-1894), was vicar of St George's, Netherfield, 1867-1887. Born in 1813 he ascended to the peerage in 1886. In 1874 he became Residentiary Canon of York Minster and Chancellor of the diocese of York. He held both these posts until his death. He died in York in 1894 and is buried in Willey, Shropshire. At this time he was Rector of Doveridge, Derbyshire (1859-1867). He was a member of a host of clergymen who founded the NBKA in 1884.

NG 23 May 1862
Bothamstall. The old saying that "A swarm of bees in May is worth a waggon load of hay" – received practical illustration in the garden of Mr Thomas Padley, farmer, on Monday the 5th May. The "oldest inhabitant" cannot remember so early a swarm of bees.

NG 16 Jun 1863
Swarming in the Town streets. Bees occasionally select the most unlikely places for swarming, but we have never heard a more remarkable instance than that which

occurred in the very heart of the old town of Nottingham on Sunday morning. Men, women, and children, have not unfrequently been selected by the queen bee as the object to which she has directed the course of her numerous followers; and the persons whose bodies have been thus appropriated have been literally covered by the winged colonists in search of a new home; but these occurrences have always been in the country.

On Sunday morning, about 6 'o' clock a fine swarm of bees suddenly made their appearances in Clayton's-yard, Bridlesmith-gate, and took possession of the branch of an old tree which has outlived the levelling tendencies of modern improvements. Mr. Clayton, baker, who occupies one of the houses in the yard, immediately procured a hive, and succeeded in sweeping the entire swarm into it numbering in all about 2,000 bees. The hive and its contents still remain in the yard; but we are glad to hear that it is the intention of Mr. Clayton to remove his strangely acquired winged labourers to the more congenial *locale* of his garden outside the town. We shall be glad to hear of this remarkable "honey fall" again.

NG 7 Aug 1863
Remarkable Increase of Bees and Weight of Stores from One Hive. Andrew Shipman, gamekeeper for Charles Keel, Esq., of Kneeton, has had the following increase from one hive which swarmed on the 3rd of June last and on the 17th of June. The hive produced a fine swarm on the 3rd of July, and last on the 12th of July showing an increase of four hives from one. The united weight of the parent hive and its offspring are 110 lbs in their hives, if the reader will allow 7lbs to be deducted from each hive (which will be quite plenty for hive bees and combs,) he will find this industrious family to have gathered the remarkable weight of 75 lbs of real store in this short space of time up to the 28th of July, an instance which has not occurred in this neighbourhood in our memory. Should anyone question these statements the owner will be happy to show them.

NG 16 Jun 1865
A Curious Swarm of Bees. On Tuesday, about noon, Mr J Froggatt, of Lenton Poplars, had a swarm of bees from an old hive in his garden, which adopted a very unusual course of exit. Mostly bees, after swarming, settle very close to the hive they have left, but in this instance the queen of the swarm took her flight towards the Trent in the direction of Wilford, accompanied by the whole cluster. A servant in Mr. Froggatt's employ followed them to the bottom of the Lenton Trent-lane, under the impression that they would settle before crossing the river, but this they did not, Ultimately he lost sight of them, and they have not since been heard of.

NG 23 Mar 1866

Nottingham and Midland Counties Industrial Exhibition. The Chairman showed to the meeting a beautiful design for a medal having the inscription and name of recipient on one side, encircled by a laurel wreath; and on the other two figures representing art and science resting against a pedestal, on which was a beehive, the emblem of industry - the arms of the four counties of Nottingham, Derby, Lincoln, and Leicester, being likewise introduced. The chairman mentioned that by the kindness of the guarantors they had a fund of £100 at their disposal, and instead of distributing it in small money prizes, it was considered better to have dies struck from the design, which had been just shewn them, and large silver and bronze medals cast.

This idea was not new. The token on the left was issued by Donald and Co in Nottingham in 1792 and was valid for use in many places throughout the country as specified. This one was to be used in Birmingham.

The token on the right was issued for competition between Nottingham schools. What this competition was is not recorded. Round the edge are is the inscription "From Labour and Industry great blessings flow" and in the centre, "Learn wisdom from above."

NG 18 May 1866

On Wednesday there was a large number of visitors to the Wesleyan Bazaar which was opened on Tuesday, in the Exchange Hall, with the view of raising funds to defray the expense of erecting the Wesleyan new chapel and schoolrooms in Arkwright-street. Mr. Simpson has added to the valuables a patent beehive, so constructed that the industrious inhabitants can be seen engaged at their labours.

BEEKEEPING IN VICTORIAN NOTTINGHAMSHIRE

The Exchange Building was demolished in 1920 to make way for the current Council House which was built in 1929. The Exchange Building was built between 1724 and 1726 replacing a shambles of buildings on the same site. It cost £2,400 at the time and comprised a four-storey, eleven bay frontage 123 feet (37 m) long.

NG 12 Jun 1868

Singular Quarrel between Bee-Masters. William Reynolds was charged with assaulting William Sharpe at Lowdham, on Friday week. Complainant said he lived near the defendant. He had a swarm of bees, and they went from witness's ground to Reynolds'. Witness asked civilly for the bees, but he refused them, and bestowed on him a good deal of abuse. Defendant knocked him down, kicked him behind, and the next day he called him a damnation thief.

Mr. Sherbrooke: How was he to give you the bees back? How could he take hold of them? Defendant: They went into one of my hives. The regular rule is to take one hive and fetch another in such a case.

A witness named Parr spoke to seeing the assault. Defendant said he told the complainant if he could get the bees out he could have them. His request was most unreasonable, and he (defendant) turned him out. He certainly kicked him, but he had a pair of light boots on. There was the greatest provocation, however. Mr. Sherbrooke concurred in the last remark, but said the defendant should not have taken the law into his own hands. He would have to pay the expenses, 21s.

NG 29 Oct 1869

John Marlow (55), labourer, was charged with stealing a beehive and bees, with about 26 lbs of honey, value 20s, the property of John Jackson, at Cuckney, on the 14th August last. A second count in the indictment charged him with receiving the same knowing it to have been stolen. Mr. Mellor conducted the prosecution, and Mr Lawrence appeared on behalf of the prisoner. The jury returned a verdict of guilty. Prisoner was then charged with a previous conviction at Worksop, but he pleaded not guilty, and it was therefore formally proved. In passing sentence the Chairman said he had been convicted without the smallest doubt in the minds of the jury or anybody else. The order of the Court was that he be imprisoned for the term of three calendar months with hard labour.

NG 10 Nov 1869
A Nottinghamshire clergyman writes that it is the custom, on the death of the master of a family, not only to inform the bees of the death, but to give them a piece of the funeral cake and beer sweetened with sugar. The bees die if not told of their master's death.

NG 28 Jan 1870
Death of a Prisoner at Southwell House of Correction. On Monday last W. Newton, Esq., coroner, held an inquest at the House of Correction, Southwell, on the body of John Barlow (sic), who died on the previous day. Deceased was 55 years of age, and in October last was convicted at the Newark County Sessions for stealing a hive of bees, being sentenced to three calendar months' imprisonment. He came from Cuckney, in this county. He appears to have suffered from bronchitis since early in December, and was placed in the infirmary under medical treatment. Mr Warwick, surgeon, said the actual cause of death was premature old age, accelerated by his prison mode of living. One of the prisoners, W. Rose, gave evidence as to the satisfaction with the treatment he had received which deceased had expressed to him. The jury returned a verdict of "Died from bronchitis."

NG 15 Apr 1870
Food for Bees. To the Editor of the Nottinghamshire Guardian. Sir, I should be glad to know if any of your correspondents are in the habit of using Tilseed Cake as food for bees, and if so, where it can it be procured, and at what price? I give below an extract from a French paper, in which it is described as a very valuable article of diet. If the facts stated in this account be true, they ought, I think, to be more generally known. Perhaps one of your numerous readers may he able to throw some light upon the subject

Extract: "The agriculturists in the Department of the Var observed one day, in the month of May last that all their bees had left their hives, although the latter were well filled and exceedingly heavy. Towards evening the bees returned heavily laden, but on the following morning set out again in a direction which was carefully noted by their masters. They at last succeeded in tracing them to a farmyard at no great distance, where cakes of Tilseed, which had been previously subjected to the press, were being beaten into a paste with water, to be used as manure for potatoes. The bees were eagerly clustering round the tubs containing the paste, evidently enjoying a luxury hitherto unknown to them. The lesson was not lost upon the owners of the bees, who immediately procured for them abundance of this food, and have now been rewarded by nearly ten times the usual quantity of produce, besides an immense increase in the reproduction of the insect." — I am, sir, yours faithfully, WBS. April 11th 1870.

Tilseed is another name for sesame (Sesamum indicum)

NG 29 Apr 1870
Singular Death. — On Friday afternoon CS Burnaby, Esq., coroner, held an inquest at the Black Boy Inn, Moorgate, Retford, on the body of George Mattashed, a man 65 years of age, who was found dead on Friday morning. It seems that deceased was addicted to drink, and on Thursday night at ten o'clock he was tipsy, and went over a wall from his yard into the garden of Mr Walker, grocer, to fetch, as he stated, a hive of bees. He did not return, and next morning was found lying dead on the footpath, having evidently died in a fit of apoplexy. The jury returned a verdict of "Died from natural causes."

NG 15 Aug 1873
Joseph Anthony, James Godber, and Jno. Price, three lads, were charged with damaging a lock and some gunpowder, the property of Thos. James Oakes at Selstone. Complainant said he had the door of his gunpowder house opened at Selstone Colliery. When witness went up he saw the defendants come out. They ran away and dropped a ring fuse and some powder behind them. The lads admitted the charge, and said they wanted the gunpowder to blow up a bees-nest. They were fined 7s 6d each.

NG 6 Nov 1874
Four Boys Convicted of Felony. At the Newark Police-court, on Thursday morning (before the Mayor and Alderman Branston), four lads named William Gardiner, Sam Davis, Jesse Millard, and Thomas Davis, aged respectively ten, twelve, and thirteen years, were charged with having, on the 11th inst, stolen three bee-hives, value £3, the property of Mr. Alfred Allen, Kings-road. The lads were all notoriously bad.

Prosecutor stated that he keeps bee-hives in a garden opposite his residence in Kings-road, and, from something said to him, he went there about seven o'clock on Tuesday morning. He then missed one hive, which had been taken entirely away, another was turned upside down, and three or four combs taken out of it A third had been taken down and put on its place again, but the comb inside had been disturbed. The damage was at least £3, as it was doubtful whether the bees would live out the winter on account of the combs having been disturbed. Afterwards Sgt. Free showed him a hive which he identified as the one which had been taken away. One hive was found in Mr Welch's stackyard, being traced by the footmarks across Mr. Frost's garden, through a gap in the hedge by the roadside, and the droppings of honey at short distances.

When the police witness compared the footmarks in the garden with some boots they corresponded exactly. One pair were plated clogs, and a portion of one of the iron rims being off there could he no doubt about them. Another pair had military heels. Sergeant Free and Constables Bullimore and Alvey proved the apprehension of the lads, and repeated the statements they made respecting each other.

The clogs and boots of the two Davises corresponded with the footmarks in the gardens. Young Gardiner and another told them Jesse Millard helped to eat the honey. The two Davises were found on a heap of gas (coke?) on the town wharf, near a fire, fast asleep, about a quarter before ten on Monday night. When apprehended they thought it was their father who had caught them, and at once said, "Father, we'll not do it any more; it was Millard and the two Gardiners who went with us to get the hives and bees."

Isaac Gardiner, 9 years of age, and brother of the prisoner of that name, was allowed to give Queen's evidence. He stated in detail how he and the four prisoners met in Castle-gate, and the particular part each took in the affair. It appeared from his evidence that, not satisfied with the honey which they joined in eating, Sam Davis got into Mr. Cope's vinery and stole some grapes, but their proceedings in that respect were hindered by the planks on which they entered tipping up. All the prisoners were stung. Millard acted as a sort of watch-man on the road, whilst the others carried out their plans.

Chief-constable Liddell gave particulars of the various offences of which the lads had been guilty, and it was stated that the School Board had been at great trouble with them. Their parents were called into court and remonstrated with. The magistrates sentenced each of the prisoners to one month's imprisonment with hard labour; Sam Davis and Millard afterwards to be sent to a reformatory for two years. The Mayor reminded them that the court had power to order them to be flogged, and if they were brought before him again he would certainly exercise that power. It was a most lamentable thing to see boys growing up hardened in crime.

BBJ 1 Jun 1876
Newark. My bees have done well during the winter, and what I have lost has been through my own imprudence, mismanagement, or want of experience. JI

BBJ 1 Aug 1876
Newark. I had my first day's slinging yesterday, and used the 'Little Wonder' with perfect success, only there might be a bigger well at the bottom, as we had to empty after each comb. I got 17 lbs. out of a hive which was a swarm only five weeks since, and two from a comb or two out of a hive which had swarmed a week since, and which, perhaps, it had been better to have let alone. Then out of an old stock which I

expected to swarm every day, I got 31 lbs., having slung every comb but the one on which the queen was. Now, will that delay their swarming? [It ought to do so. Ed.] I could not find a queen-cell, but if I had another empty hive, which I have not, I should take her out and divide the hive, as I have lots of queen-cells on combs in other hives. Our flower show on Thursday was not well attended, but the bees got their share of attention. Every one was very pleased.

The "Little Wonder" was a popular rotary honey extractor at the time. It was made by Abbots, Southall, London, a well-known manufacturer of beekeeping equipment

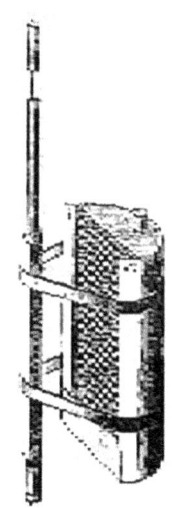

BBJ 1 May 1876 As may be seen, the cage and can revolve round a shaft, the bottom end of which is furnished with a short spike, and the top with a long iron pin, which turns in a wooden handle; now, if an unsealed comb be placed against the wire-work, the bottom spike pressed perpendicularly on to a piece of board, on which the operator should stand, and with either hand the handle at the top be made to describe a small circle, the machine will begin to revolve, and with five minutes' practice may be made to describe from 130 to 200 revolutions per minute with very slight exertion. This, of course, causes the liquid honey to leave the cells on one side of the comb, when, still holding the machine with one hand, the comb is reversed with the other, and in another minute it will be found empty.

BBJ 1 Jul 1877
It is suggested that a Bee and Honey Show might be organised with Gedling and Carlton Flower Show near Nottingham if a few friends to the cause in the district would combine for the purpose. Mr WC Cheshire of Gedling will be very glad to act as Hon. Sec *pro tem*, and will give any information he may possess to intending exhibitors of hives, etc.

William Culley Cheshire was baptised on 2nd August 1848 in Gedling. He was the second son of John and Hannah Ann Culley who were married at St. Peter's Church, Nottingham. In the 1881 Census his occupation was given as Nurseryman. He was to serve the Association as Secretary for many years. He lived in Maltmill Cottages, Bulwell at one time.

A search of reports of all the local flower shows for the next two years indicated that none of them had honey shows included.

Shall the Lincolnshire BBKA extend its borders over the whole of the Midland Counties, and also of Yorkshire and Lancashire, and so constitute one grand Central Association? I venture to suggest as a question for those at the helm of the

Lincolnshire Association to weigh over. Myself a bee-keeper residing beyond its present borders, sincerely hope they will give the matter due consideration. From my own observations, as also of others, of the thoroughly spirited way the Lincolnshire have hitherto carried forward the work, is a convincing proof of its determination to achieve the objects for which the Association was, at the outset, instituted. Perhaps if a little pressure were brought to bear upon the present Committee, they would take the matter in hand, that done, those of us who have had the opportunity of witnessing past doings of the Lincolnshire may contentedly wait the results. I would beg all our leading men in the above-named committee to at once throw their weight into the subject. We should then, I believe, accomplish what (to my mind) would prove in every way most beneficial. A Notts Bee-Keeper December 20, 1877. (This communication although dated 20th ult. Arrived only on the 27th. Ed. BBJ)

BBJ 1 Jan 1878
Charles Nash Abbott, editor of BBJ suggested that, in addition to the national organisation, a Central Association should be formed with its base in Grantham. This town was chosen due to its position in relation to many others places including Nottingham. To stop complaints that Grantham was too far to travel Abbott pointed out that exhibitors travelled from Scotland to take part in honey shows held in London.

"All cottagers adopting the 'humane system' of bee culture (not killing their bees using buring sulphur) shall be honorary members (to be certified by the clergy of the parish)". People with means would be expected to make donations of up to 10/-.

NEP 15 May 1878
Newark Show. Mr John Elam (?), Basketmaker of Newark showed his skeps.

BBJ 1 Jul 1878
John Cree, Balderton, Newark. Bees are sufficiently numerous to hang outside the hive can scarcely be accounted queenless. Queenless stocks generally dwindle away through the death of the bees, and there being no young ones to take their places. They are probably hanging out waiting for better weather, to swarm; but if like last year, better weather does not come, they will do as they did then, and be profitless. Why not swarm them artificially and set them to work? If you fear to undertake the task, perhaps some Newark beekeeper, who is not afraid, will volunteer to help you.

John Cree was an ironmoulder born in 1850. He married twice, both his wives being called Elizabeth. His grandson John Dennis Cree lived at Willow Cottage in Balderton where John Cree had moved after his second marriage in 1886. He died in 1932 in Newark.

NG 23 Aug 1878
Lincolnshire BKA. The third annual exhibition of honey, bees, hives, etc., of the above association, was held in the Corn Exchange, Stamford, on Tuesday last. The honey

classes formed a grand and striking show. Mr. Measures, of Upton, Southwell and Mr. Sandaver, of Southwell, were exhibitors.

Samuel Sandaver was a consistent competitor and judge in the various horticultural shows in the Southwell area. He was born in Southwell in 1836. In the 1881 Census he is shown as a Gardener and General dealer and lived in Southwell with his wife, Sarah (b.1836 Southwell) and his son James W (born in Southwell in 1870). James was a scholar at the time.

EC Walton and Co. Sutton-on-Trent

It all started with bees! In 1878, Mr. EC Walton was a lecturer of beekeeping in Nottingham and lived in Muskham, near Newark. Due to the high price of carpentry his students were often left without a beehive to work with so he decided to set up a manufactory in his home village to help support his student's enthusiasm of beekeeping, and called it Walton's Muskham Beehive Works. It didn't stop at beehives. Local farmers, fed up with their own collapsible efforts at providing shelter for their poultry, came to him for chicken-coops. *Walton's were building garages by the time Henry Ford was building the model T!*

Summer-houses came next, or "consumption shelters" as the Victorians thought that fresh air was the best cure for the peculiar Victorian disease of consumption.

Prices for his products ranged from two shillings for a modest chicken-coop to over a guinea for a particularly ornate Consumption Shelter. Waltons Muskham Beehive Works was doing very good business but it wasn't making very many beehives. So by 1900, the company's name had changed to E.C. Walton & Co.

The company was starting to win some of the many medals it has collected over the years for excellence in manufacture, and started to expand during the early years of the 20th Century. This period also saw the arrival of EC Walton's son, EDL Walton who took over his father's role and built the company's new factory at Sutton-on Trent, in 1927, where the factory still resides today (2018).

NG 14 Feb 1879

Proposed Beekeepers' Association. To the Editor of the Nottinghamshire Guardian. Sir, Would you allow me a little space in your columns to call the attention of beekeepers and others interested in the habits of the bee, as well as the consumers of honey, to the fact that there are several in the county desirous of forming an association with a view of holding an annual exhibition of bees and honey in Nottingham. The assistance of clergymen and gentlemen is earnestly invited in order to give the movement a fair start. When it is known that in Nottingham and its neighbourhood there are hundreds who keep bees, it is a matter of surprise that an association has not long ago been established. The society will have for its object the better management of bees, and will bring before its members the most economical and profitable method of culture. Nottinghamshire has been behind its neighbours in the want of an association of this kind. Lincolnshire has long boasted of a flourishing society, and its exhibition at Grantham last year afforded a rich treat to the many lovers of bees in this county who visited it. Wolverhampton, Westbury, Devon, and Exeter, and many other towns which I could mention, have long ago established societies which have been beneficial both to bee-keepers and consumers of honey. Those who visited the Crystal Palace and Alexandra shows last season express themselves highly pleased with what they saw, and the profit they derived from witnessing the numerous methods now in existence for the better management of bees and the production of purer honey. All that is wanted to form a successful society in Nottingham and county is the hearty co-operation of clergymen, gentlemen, and others interested in the matter, and I shall be glad to receive the names of such with a view to arranging a place of meeting for taking into consideration the best means of forming an association. — Yours, etc.

WC Cheshire. Maltmill Cottages, Bulwell, Feb. 8th, 1879.

NG 14th Mar 1879

Proposed Bee-keepers' Association. On Saturday afternoon a meeting was held in the Mechanics' Hall, Nottingham, to consider the best means of forming an Aparian Association for Nottinghamshire, with a view to an annual show, and lectures on the best mode of bee management and culture. Capt. Rolleston was expected to be present, but owing to another engagement did not attend, and the Rev. Canon Smith, of Southwell, was voted to the chair.

Mr. Cheshire, of Bulwell, said he was sorry that Capt. Rolleston was not able to attend, but he was sure that whatever position they placed him in he would do his best towards promoting the objects of the society. He (Mr. Cheshire) knew thirty other gentlemen who were willing to assist the society in any way, and he was sure there was nothing to fear against its becoming a perfect success if they would render aid in the best way they could.

Mr. Joseph Phipps, of Arnold, said doubtless they were all aware of the purpose that had brought them together, and that was to abolish killing the bee for the sake of the honey, which was greatly practiced in this neighbourhood. Thirty years ago he had done the same thing, but recently had adopted other plans, and by then means he had been able to gather from 25 lbs to 50 lbs of honey in a season, without destroying many of the bees. The society was anxious to enlist the attention of the working classes, and also to induce cottagers to become keepers of the bee; but there were some persons who thought that because there were 20 stocks of bees in a village, it was quite sufficient to gather all the honey there was. The society could not be formed without incurring some expense, but he would be glad to contribute according to his means towards carrying out the object in view. The society intended during the summer to exhibit it the Horticultural Show, London, several hives of bees and hives constructed on different principles.

The Chairman thought that, if possible, something definite should be done, and a resolution put before the meeting either to adjourn it or form a committee. Mr Woodhouse then proposed that a committee consisting of a chairman and twelve members should be formed, which was seconded and carried. Mr. Talbot remarked that about £400,000 was being paid yearly to foreigners for the importation of honey into England, and there was no doubt that if cottagers were to cultivate bees as one profitable system they would be able to obtain as much honey and wax as would pay their rent.

The following committee was then formed: Captain Rolleston, Messrs. Phipps, Clark, Woodhouse, Talbot, Parry, Cheshire, Ward, Varley, Smith, and Adams, with power to add to their number. It was proposed by Mr Woodhouse, and seconded by Mr. Sissling, that Mr. Cheshire be elected secretary, and this was carried unanimously. A vote of thanks to the chairman for presiding concluded the proceedings.

Col. Lancelot Rolleston, DSO, b. 1847 Greasley; d. 1941 at Watnall Hall. This was built about 1690 and was demolished in 1962.

The Rolleston Window in Greasley St. Mary's church was dedicated in 1960 to the memory of Lancelot Rolleston, KCB, DSO 1848-1941 and his wife, Emma Maud Charlotte. The window poignantly incorporates coloured glass taken from Watnall Hall and an illustration of the family home from around 1690.

"..the old-fashioned garden, with its thatched bee-house, full of murmurous sound. Of that bee-house I made a closer inspection. It was designed by Mr. Rolleston, who takes considerable interest in bee culture, and is large enough inside for a study. As a matter of fact, Mr. Rolleston occasionally shares this house with the bees. The boxes in which the insects deposit their honey are so arranged that they cannot fly about in the interior of the apartment, whilst their operations can be watched and studied through glass. So the bee-house serves a double purpose, besides being an ornament to the garden. Whilst the bees are making their honey, and arranging their domestic matters in the glass cases, Mr. Rolleston is writing his letters at a table, and the apartment is filled with a soothing sound."

Leonard Jackson 1881

Joseph Phipps (1817-1880) was a schoolmaster in Arnold. Born in Nottinghamshire he married Kate (Kitty) who died in February 1882 aged 63. She was born in Leicestershire. In 1855 he gave a talk on electricity to the people of Papplewick at their request. About 1858 he had set up the Arnold Academy whose choir in 1871 sang at the funeral of the Duchess of St. Albans. He was on the platform when a new school was opened in 1868. In 1877 he was living at Osborne Villas, Arnold. After his retirement as a Schoolmaster he was appointed a Registrar of Births, Marriages and Deaths to the Basford Board of Guardians. Records of burials in Arnold Churchyard include "Joseph Phipps, Schoolmaster, Dec. 13, 1882, aged 63." From the note below this is obviously wrong. It should read "Kate Phipps, relict of, etc."

On Saturday, 11th December 1880 an inquest was held at the Greyhound Inn, Arnold by

BEEKEEPING IN VICTORIAN NOTTINGHAMSHIRE

Mr. D. Whittingham, district coroner, upon the body of Mr. Joseph Phipps, 63 years of age, who committed suicide by hanging on Friday. The only witness called was Mrs. Phipps, who said she was the wife of the deceased who was a retired schoolmaster. She found him on Friday in his bedroom hanging behind the door, about a quarter-past one o'clock. He had been in bed all the morning, and witness was going to see him. He was dead, hanging by a strap. Deceased appeared a little excited in the morning. Witness thought that deceased was troubled about his elder brother, who had been ill for some time, and was obliged to go to the asylum a week ago. He said on Friday that his poor sister (Ann?) had not lived to see her brother taken to the asylum. Certain marks which were upon deceased's face had been caused by his falling on Thursday night He was somewhat the worse for drink when he came home on that night and had eaten very little during the week. — A verdict was returned that deceased committed suicide by hanging himself, being at the time in an unsound state of mind. His elder brother died four days later.

NG 21 Mar 1879

The Beekeepers' Association. To the Editor Nottinghamshire Guardian. Sir, I do not remember the time when I felt more delighted than when I had the pleasure of reading in your paper an account of a meeting which was convened at the Mechanics' Institute on Saturday last, for the purpose of considering the best method of leading the working classes into the study and culture of that long-neglected and cruelly treated little English labourer, the working bee. May I be allowed to say, through the medium of your widely-spread paper, the Guardian, for the profit and pleasure of my fellow working men, that for many years past I have been a beekeeper, and the large majority of apiarians have been very particular in preparing the brimstone rag and slit standards for the speedy destruction of my poor little friends, who, while they lived through the short span of summer life allotted to them were constantly toiling night and day for me; but, thanks to those who have studied the habits of our little friends, that cruel work has now, as regards the practice in taking the stores of delicious nectar which they so freely lay up for me and mine, for a long time ceased with me, I do sincerely hope that the clergy and ladies and gentlemen throughout this county will give some portion of their time towards inducing all beekeepers in their respective neighbourhoods to join the Nottingham Town and County Bee-keepers' Association as soon as possible, so that they may have the advantage of receiving instruction from those who thoroughly understand the habits of the British honey bee, and the best means of reaping the greatest amount of profit and pleasure which can to almost every season he enjoyed by those who are willing to act on the humane system which is now practiced by some of the most intelligent apiarians not only in this, but in most parts of Europe and America. — I am, sir, etc.

                        An Old Bee-keeper. Arnold, 13th March, 1879.

David Charles Woodhouse was a timber merchant with premises on Canal Square Wharf. His timber business failed in July 1884 due to the increased importation of timber from abroad.

Rev Canon Robert Frederick Smith, MA, was a Minor Canon of Southwell Minster. A Minor Canon is a member of staff on the establishment of a cathedral or a collegiate church. A number of religious songs were published by him. He was the Chairman of the meeting which brought into being NBKA. He was previously curate and then vicar of Halam (1861- until he died in 1905), which service is commemorated inside the church, and brother of Thomas Woollen Smith, vicar of Calverton.

On 24th May, 1877, he married Valentia, younger daughter of W Hodman Donne, Esq. of Portland Place and Mattishall, Norfolk, JP and DL of the county. The ceremony was performed by the Very Reverend Dean of Lincoln assisted by the bridegroom's brother. Valentia died in 1917 in Southwell at the age of 78 years.

Mr. Donne's youngest daughter Valentia was married on 24th May to Rev. RF Smith, Minor Canon of Southwell Minster.

NG 28 Mar 1879
NBKA. On Saturday afternoon a meeting of this newly-formed association was held in a room in the Mechanics' Institute. The Rev. Canon Hole, vicar of Caunton, was voted to the chair, and there was a very fair attendance. The Secretary (Mr WC Cheshire) read letters from Mr. Herbert B. Peel, secretary to the British Bee-keepers' Society, containing some useful information. He also offered to send down to Nottingham a lecturer at the expense of the London association, to deliver a lecture on some subjects connected with bee-keeping. In case a county association was formed, the British or Central Association could offer them a silver and gold medal for competition amongst the members, together with the use of their bee-tent for the annual show at a nominal charge. Each county association that expected to receive the benefits must first affiliate itself to the Central Association by a payment of one guinea per annum. Since the meeting of the association in London on the 12th of February, county associations had been set on foot in Surrey, Kent and Suffolk, and Mr. Peel thought they should soon hear of new associations in Buckinghamshire, Warwickshire, and Gloucestershire. The object was fully discussed at the meeting on Saturday, and it was resolved to take the steps necessary for arranging a series of lectures. A silver and gold medal were also offered for competition at an annual exhibition. The following were elected on the committee: — Rev. Canon Hole, Mrs. Hole, Mrs. AM Francklin, Rev. FW Smith, and Messrs. Phipps, Woodhouse, Talbot, Perry, Ward, DC Woodhouse, Bail, Wheatley, Milner, Allsebrook, Cheshire, and Sissling. Mr. Phipps was selected treasurer.

The Hon Mrs. Alice M Francklin of Gonalston was the daughter of Lord St. Vincent, and the wife of John, an ex-master of foxhounds. "One of the right sort." She was born in 1849 at

BEEKEEPING IN VICTORIAN NOTTINGHAMSHIRE

Witherington, Cheshire.

Rev FW Smith could be Rev TW Smith. This would be Thomas Woollen (sometimes Woolen) Smith, vicar of St. Wilfrid's, Calverton (1874 – 1905). In the east window of St. Wilfrid's he and his wife are commemorated. He was responsible for the restoration of the church buildings 1881. Previously in 1874 he had given a new altar, the old one being used to make the cross which stands over a beam above the church steps.

Samuel Reynolds Hole, (1819-1904), curate and vicar of Caunton from 1844-77, rural dean of Southwell 1873-87, proctor in Convocation 1883-7 and Dean of Rochester from 1884 until his death, was a fine example of a high-minded, genial, hard-working parish priest, of whose mermory Nottinghamshire will always be proud. He was quite besotted with roses, loving them nearly as he loved fox-hunting. In Caunton, he grew 400 varieties and later when appointed Dean of Rochester, 135 roses filled the deanary garden. At Caunton he instituted daily services and never omitted a daily visit to the village school, but his clerical duties were varied by hunting, shooting, and other rural sports, and he was an enthusiastic gardener.

After the death of his father in 1868 he was squire of Caunton as well as vicar, and his genial humour made him popular with all ranks. Hole married, on 23 May 1861, Caroline, eldest daughter of John Liell Francklin of Gonalston. He is buried at Caunton.

William Pole Jones Allsebrook, JP. Born in Yoxall, Staffs in 1835, died 1906 in Cork on a visit to one of his sons, and is buried in Wollaton churchyard. He lived at Old Park Farm in Wollaton where he was a tenant farmer under Lord Middleton and a Valuer.

NEP 11 Sep 1906
At a meeting of the Basford Rural District Council to-day, the Chairman made sympathetic allusion to the death of Mr. Allsebrook. He was member of the original rural sanitary authority and the old Highway Board, of which for considerable period he was chairman. He was for many years associated with the Council.

The Annual Soirée of the Naturalists Society took place on the 11th of March 1879 in the Mechanics Hall, Burton Street. Mention of Mr WC Cheshire's contribution must not be omitted, as the improved and observatory hives, showing bees at work, virgin honey in the combs, and specimens of bee-bread, with improved bee furniture were among the most entertaining exhibits in the hall.

NG 11 Apr 1879
We pay a very considerable sum in hard cash to foreigners for honey which might just as well go into the pockets of our own countrymen, and would go into them if they

bestirred themselves. Old methods of working will not, however, do now-a-days. Our ingenious American cousins have invented all sorts of appliances, and our home bee-keepers have added to them, so that the old smothering of the swarm and so decreasing the stock, together with the old bad methods of extracting the honey from the comb need no longer be followed. If any are anxious to know all about the honey crop and how to get it the means lie ready to their hand. There exists an association called the "Nottingham and Nottinghamshire Bee-Keepers' Association", of which any one can, by paying an annual subscription, become a member, and the members and officials will gladly give all information and aid in the disposal of the honey when gathered. Mr. LR Rolleston is president, the Rev. Canon Forester, vice-president, and amongst the committee we note the well-known name of Canon Hole. The secretary is Mr W C. Cheshire, Gedling. The society will, we believe, hold a meeting this afternoon, which all interested in the subject would do well to attend. Clergymen and land-owners would confer no small boon on their poorer neighbours if they can introduce the practice of keeping bees. We should thus be made independent of foreign supplies, and the earnings of the cottager's household would be increased.

NG 11 Apr 1879

A meeting of the members of NBKA, under the presidency of the Rev. TW Smith, vicar of Calverton, was held in the Mechanics' Hall, on Saturday afternoon. The business was principally of a routine character, the meeting having been called for the purpose of revising the code of rules which had been framed for the management of the association.

The minutes of the last meeting having been read and confirmed, the Rev. Chairman called attention to the circular which had been issued to all interested in the prosperity of the society. The rules were then gone through seriatim, the Chairman inviting suggestions or discussion upon any points which in the opinion of those present would prove more advantageous to the association.

Rule 5, which states that the minimum subscription of members shall be 2s 6d, honorary members 10s 6d, and that subscribers of 10s 6d shall be entitled to a family ticket to the exhibition, gave rise to a difference of opinion. A gentleman moved as an amendment that a medium scale of 5s, entitling the holder to a family ticket to admit three, should be adopted, believing that the finances of the society would thereby be enhanced. The amendment having been seconded and put to the meeting, it was found that the majority were in favour of the rule as it at present stood.

None of the other regulations having been challenged they were, with a few slight alterations and additions, ordered to be printed and circulated. Several gentlemen paid their subscription money, and were entered on the books as members. A vote of thanks to the Rev. Chairman brought the proceedings to a close. The secretary (Mr.

BEEKEEPING IN VICTORIAN NOTTINGHAMSHIRE

Cheshire) had on view a hive constructed on novel and most economical principles, peculiarly adapted to the requirements of cottagers and amateurs.

28 Apr 1879. (History of Gedling, 1908 by Charles Gerring)
At the annual cricket club dinner in Gedling this verse was sung (amongst others):
> You've a gentleman keeps the "Chesterfield Arms"
> He's a fine lot of bees and often has swarms.
> One hot summer morning the bees would not rest
> So he'd three swarms before dinner and one very good cast.
> Chorus:
> Up and down Gedling you can walk about free
> And hear the new clock strike its one, two, three.

NG 9 May 1879
A general meeting of the members of NBKA was held in the Mechanics' Hall on Saturday afternoon. There was but a small attendance, and Mr. Ashworth occupied the chair. A discussion arose as to whether it would be to the advantage of the society to join it with the Naturalists' Society. One member was of opinion that if this course was adopted they would not have so many members as at present. They were formed into an association for a specific purpose, and if they affiliated themselves with the Naturalists' Society there would be a double subscription to pay, which would prevent the poorer classes from becoming members. The subject was dropped without any resolution being come to. It was proposed, seconded, and carried that a show should be held during the autumn, Mr J Phipps stating that this would be a means of bringing the objects of the society before the public. After a resolution had been passed empowering the secretary, Mr. Cheshire, to collect subscriptions the proceedings terminated.

This could be Rev. James Rhodes Ashworth, chaplain to Mr William Frederick Webb (African explorer) who bought Newstead Abbey in 1861 from Colonel Wildman's widow. The Wildman's outbid Queen Victoria in buying the estate. James was born in 1851 in Manchester and in 1881 he was living with his wife, Florence HC (born in Italy in 1850) and their children in Parsonage House.

BBJ 1 Jun 1879
County Associations. An Association has been lately formed for Nottinghamshire, and we notice that the Rev. Canon Mole *(sic)*, who is so well known in connexion with rose-growing, is one of the Committee for that Association. We wish it every success. We are also informed that steps will shortly be taken under influential auspices

NG 20 Jun 1879
Nottingham and Notts Bee-Keepers' Association. A meeting of the members of this

association was held on Saturday afternoon in the Mechanics' Hall. There was only a moderate attendance. The chair was occupied by the Rev. TW Smith, of Calverton. At a previous meeting it was decided to hold a show in the Arboretum during the autumn, and a discussion now arose as to what days would be the most appropriate. The Chairman remarked that it was thought desirable by the committee to hold the show either in the first week in August or September, but several members were, however, of opinion that owing to the late season August would be too early. The secretary, Mr Cheshire, said from what he had observed he thought there would, this year, be a plentiful supply of honey which would be of better quality than usual. After some further discussion, it was proposed by Mr Holmes and seconded by Mr J. Phipps, that the show be held on the Friday and Saturday of the first week in September. This was carried unanimously.

The Secretary remarked that he had written to several gentlemen asking them to collect a few honorary contributions but he had not received any answers. There was at the present time about £13 18s in hand, including three guineas from Captain Rolleston, one guinea each from Mrs Musters, Canon Forester, and Mr J. Robinson; and 10s 6d from Captain W. Sherbrooke. The subscriptions promised, which had not been collected, would bring the amount up to about £20.

The Chairman thought it would be desirable to write to Mr. H. Peel, of the BBKA, to advise a scheme for the first meeting of the society. The Secretary said he had already done so, and Mr. Peel could recommend nothing better than their following what had been done at the first meeting of bee-keepers' in Herefordshire. He (the secretary) would, by the next meeting, draw up a schedule of the amount given in prizes at different meetings held in England for some time past, and from this they would be able to form some idea as to what they should give as prize money. Several new members were enrolled, and the proceedings terminated with the usual vote of thanks to the chairman for presiding.

The Secretary would have known that there was, within the county, a group of the landed gentry and clergymen who regularly met for various social, sporting and other events. Many marriages took place between the offspring of families within this group. Fox hunting was the most common theme with the South Notts Hunt (with Mr Liell Francklin as Master) most prominent but other members had their own packs of hounds. There was also a strong link with the Conservative party and Freemasonry. The guest lists for balls, weddings, funerals, etc. always have a number of the same names. These people also employed a large number of "cottagers" on their estates so it would be in their interest that the lot of these tenants would be improved.

Capt William Sherbrooke JP. RN. Born in Calverton. He married Margaret McDonald Graham in Renfrew in 1870. He retired from the navy with the rank of Commander and is recorded in the 1881 Census living in Colwick Hall as head of the household but by 1894 he had moved to Oxton.

BEEKEEPING IN VICTORIAN NOTTINGHAMSHIRE

The Arboretum was opened on the 11th May 1852 by the Mayor, Mr W Felkin and the Sheriff of the Borough, Mr Ball, in front of 30,000 people. The layout and design was carried out under the supervision of Samuel Curtis, a botanist and horticultural publicist in 1850 who had previously been involved with the layout of Victoria Park in the East End of London in 1842. The main aim of the design for Arboretum was to take advantage of the landscape setting whilst providing an interlinking network of walkways and socialising areas. The plantings were laid out in what is known as 'The Natural Order' to provide an educational link to nature through botanical interpretation.

William Felkin was Mayor of Nottingham 1850-2 and T Ball was Sheriff 1851-2

NG 27 Jun 1879
One or Two Things. 1. Bees and Scarlet Runners. Bees are coming forward. Bees are making themselves felt (they always do, if you poke them up enough), but Bees are uniting. {Nothing new in that; of course they do, when they swarm.) They are Trade-unionising. They have now a county, and a town and county association of their masters — or rather, owners; for there are few people, I take it that are masters of Bees, if Bees have a mind otherwise. But it is now a fact that there is the Nottingham and Notts. Bee-keepers' Association; and they have had a meeting, and have decided to hold a show of Bees some time in August or September in the Arboretum. So that, you see, I am quite right when I say that, locally, Bees are coming to the front. It was the reading of the account of the meeting of the Bee-keepers on Saturday that set me athinking about Bees. There seemed to come a buzzing about my ears which I could not at once get rid of; and, as I am deeply interested in Bees (every gardener is!) I thought them and their doings over right well, and at last my thoughts settled down to one thing in connection with them that has exercised my mind, less or more, ever since the idea was conveyed to it.

Some time in November last Professor Henslow communicated an article to our contemporary, the *Gardeners' Chronicle,* which I dare say rather astonished many more gardeners besides myself. The communication was on the barrenness of the Scarlet Banner Bean. The leaned Professor said that the Scarlet Banner Bean, different from all the Dwarf Beans, was entirely dependent on Bees for the fertilisation of its flowers. He said that, physically, the construction of the flower entirely, or very nearly so, precludes fertilisation by its own agencies, that there must be the admission of insects to carry from the stamens the fructifying pollen to the pistil, in order to make the flowers fertile. He said, also, that this is so in the Scarlet Runners' native country of Mexico just as much as here; and, in his opinion, this is it that accounts for the many barren flowers that there are at times with us, which fall off instead of filling out into Beans.

And he went on to say that though we hear the hum of the Bees about the Scarlet Runners it does not follow that the flowers will get fertilised; because both the bumble and the hive Bee, disliking the trouble and labour of entering each flower at its opening, soon find out that they can make a hole at the bottom of the tube and extract the nectar that the flower contains without going at all near the stamens and pistil; hence, of course, barren flowers, to our great astonishment; for, as we know, the Bees have been about them, and we are lost in wonder how it is that the flowers fall off without coming to perfection. Dwarf Beans, the Professor says, are self-fertilisers; which is lucky for gardeners, or how on earth should we go on in the forcing season, when the Bees are all dormant?

All this lands us in a fog, and I am afraid it leaves us there. If this is so, and, coming from such a source, it is worthy of respectful and thoughtful attention, what are we to do? It appears to me that when we discover that it is so — if it is so — the number of Bee-keepers will be increased, and the ranks of the Bee-keepers' Association strengthened. I sound this note so that everybody may be on the look-out. I may as well say that, personally, I know nothing about the matter. Like Goliath's astonishment about David's stone, such a thing had never entered his head before, so with me; I really had never connected Bees so closely with Scarlet Runners as that, and certainly never concluded that they were indispensably necessary to their fruitfulness. I shall look out now; the Bees will be watched. I mean to know all about them and their ways; and so, ye Bees lookout! I send out this note that all those interested may assist to the best of their power.

BBJ 1 Jul 1879
Notts, May 27th. Bees are not behind the vegetation here, everything being late. Sycamores are out, but not apples, and (weather permitting) some hives are ready for supering.

NG 4 Jul 1879
Advertisement - NBKA will hold their next general meeting in the Mechanics Hall on July 5th at 2pm when Prize-List will be arranged. Anyone wishing to make a proposition for the benefit of the Cottager Classes should attend. Subscriptions due, and members solicited. WC Cheshire.

BEEKEEPING IN VICTORIAN NOTTINGHAMSHIRE

NG 11 Jul 1879

Nottingham and Nottinghamshire Bee-Keepers' Association. A general meeting of the members of the above association was held on Saturday afternoon in the Mechanics' Hall for the purpose of drawing up a schedule of prizes for the forthcoming show. The Rev. TW Smith, of Calverton, occupied the chair.

A letter received from Mr. H. Peel, secretary of the BBKA, in which was enclosed a schedule of the show held by the Herefordshire BKA, was read by the Chairman, and after some discussion it was decided to adopt the schedule as it had been received, the society offering prizes amounting to about £17 10s.

A stall will be provided at the show for the sale of honey. Besides the prizes given by the society, the BBKA offer a silver medal, bronze medal, and a certificate to the three members of the society who are awarded the largest amount of prizes in the honey class. Members of the society will be charged 1s for each entry, and the non-members 2s 6d. The first eight classes are open to all England. It was resolved to forward a guinea to the secretary of the BBKA as an affiliation fee to that association; and after other business had been transacted the proceedings terminated.

Bees. To the Editor of the Nottinghamshire Guardian. Sir, I noticed the other day an article in your columns respecting our favourites both as a hobby and an object of industry. Your correspondent says that for thousands of years bees have been kept for their honey, but that at the present day they are only a hobby. Allow me, sir, to express my opinion that bees are as much kept for their honey now as ever they were and ask why they should not be both a profit and a hobby at the same time.

With regard to your correspondent's remarks, I think that the enemies to bees are very few when they are properly cared for. The tom-tit, sparrow, flycatcher, robin and even the linnet are mentioned, but during my experience, since 1865, I have not had any serious losses at all.

The mouse is also mentioned. It is certainly an audacious enemy in winter. It crawls through the entrance to the hive, or gains admittance in any way it can, devours the honey and chumbles the comb, but this is owing to the neglect on the bee-keeper's part. If he secured the entrance when the bees are not on guard he would have no more trouble from the mouse.

As regards the snail, the same precautions may be adopted as with the mouse, but even if a snail does get inside the hive they do very little harm. Before it has been there very long it is securely fastened down with wax, so that it cannot move.

The greatest enemy to the bee is the moth, and it requires all the care of the keeper to keep this pest down. It gets in any crevices there may be, in both old and new hives. It will lay eggs outside in any niche, or inside if it can gain an entrance, in the cells. I have seen hives eaten, and quite spoiled by this pest. It should be eradicated whenever it can be seen. In the autumn it is a good plan to put sticks covered with treacle or birdlime, inside the hives, and you can thus catch from a dozen to a score of moths from 8 pm to 6 pm.

The spider is only an outside nuisance. Its web gets on the wings and legs of the bee and causes great annoyance.

Neat and well-made hives, and covers, placed in the open air by the side of the path, and two examinations during the season, will do away with this nuisance. Wasps attack the hives in the autumn when there is a scarcity of food. As many as possible should be killed in the spring. The entrance of the hive should be made smaller in the autumn, and in this way the strong hive will take no harm.

With regard to the distance a bee will travel for honey and bee bread, my experience is that it will not go out of a radius of two and a half miles. It is of no use to dust your bees when they turn out in a morning. If you watch a populous hive and dust the bees with any particular colour, you will find the bees returning with pollen from many differently coloured flowers. I have often wondered in spring from what flowers my bees have coloured themselves. I have seen seven come to the entrance of the hive in one minute, all with different shades. Feeding on artificial pollen is one of the secrets of bee-keeping in the spring. The bees should be fed on the top of the hive, and not at the bottom. By so doing, you will keep your stock in a proper position, keeping up the temperature of the hive, which in the breeding time is so important.

Your co-respondent writes that he does not want to create a discussion. I think a discussion would be a good thing. Your correspondent is probably aware that there it an association to be formed in Nottingham called the "Nottingham and Notts Bee-Keepers Association" and that a show will be held in the first week in September of bees, hives, and all kinds of bee furniture. Manipulation of hived bees will take place, and every description of bee produce will be on sale. I hope I shall have the pleasure of meeting your correspondent on that occasion. I am, sir, Ac, WCC. (WC Cheshire?)

Death of a Child at Arnold. On Friday last, a child, fifteen months' old, named Blatherwick, living with its parents at Arnold, died under rather singular circumstances. A few days ago whilst the child was playing in the street it was stung by an insect supposed to be a wasp or a bee. No notice was taken of the matter until the child's leg began to swell, and the mother on examining the leg extracted the insect's sting. The child, however, became worse, whereupon Dr. Beid's assistant was sent for. He treated the child for inflammation, but in spite of all medical aid the child died on Friday from the effects of the sting.

BBJ 1 Aug 1879
Nottingham and Notts Bee Association,
The Association will hold their exhibitions of honey, Bees, Hives, &c, and Practical Apiarian Manipulations, during 1879, in the Arboretum, on Friday and Saturday, September 5th and 6th. Classes 1 to 8 open to all England. Classes 9, 10, 11, 12, 13, and the Honey Fair, open to members of the Nottingham & Notts. Association, and other residents in the County of Notts.

BBJ 1 Aug 1879
A general meeting of the members of the BBKA, held in the Albert Hall, London, on Wednesday, July 23, at 6pm. The Secretary gave a list of the County Associations which had already been established, viz. in Shropshire, Lincolnshire, Hertfordshire, Devon, Dorset, Surrey, and announced that steps were being taken to form a similar association in Nottinghamshire, and spoke of the mutual benefit which the Parent and County Associations would receive from one another.

BBJ 1 Aug 1879
Nottingham, July 24th. Our bees are in a poor way here. Feeding has been necessary all the summer. I left off feeding one of my stocks a few days ago and the result was that they began to cast out white bees and drones. I am told that up to the present week we have only had twelve days without rain during the whole of this year, a statement I can well believe. I see some of your correspondents say they have had more swarms this season than they knew what to do with. That is not our case in this district, as I know several beekeepers who have had no swarms at all. We are to have a show at Nottingham during the first week in September but, as far as honey is concerned, I should imagine the exhibits would be nil. WS

NG 20 Aug 1879
Wanted, 20 Stocks of bees in common straw skeps; also, all Condemned Bees, at per lb, and any quantity of clean worker comb. Send price and particulars. —Address Secretary, Nottingham and Notts. Bee Association, 20, Mechanics' Institute.

NG 25 Aug 1879
Nottingham and Notts. Beekeepers' Association. One of the most animated meetings in connection with this newly-formed society took place on Saturday afternoon in the Mechanics' Hall, under the presidency of Captain Rolleston. The attendance was larger than at any previous gathering, and the interest evinced by the members present augured well for the success of the forthcoming exhibition, which will take place on Friday and Saturday, September, 5th and 6th. It was resolved, as there

would naturally be a paucity of honey from this season's hives in consequence of the unpropitious weather, that honey from last year's stocks be also exhibited. This proposition met with general approval, and several bee-keepers intimated their intention of forwarding a supply in order to make the exhibition as attractive as possible. Despite the adverse season many good stocks will be on view. Numerous hives, constructed on the most approved principles, from London and other places, will be exhibited.

The present year will be one long remembered by bee-keepers, as many have had entire stocks destroyed by the cold weather, whilst others have had to feed their bees with an abundant supply of sugar in order to keep them alive. The manipulation of bees will form an interesting feature of the show, and those uninitiated in the process will doubtless derive much pleasure from witnessing the operation. It was decided that the services of four gentlemen as judges should be obtained, instead of two as hitherto arranged.

At the suggestion of the chairman, it was decided that a band should be engaged for the two days in order to enliven the proceedings. The members present appeared to take great interest in the proceedings and intimated their intention of doing all they could to make this, their first exhibition, a success. No doubt the novelty of a bee show in Nottingham, together with an exhibition of the most approved hives and appliances, will bring together a large number of visitors. A vote of thanks to Captain Rolleston for the interest he has taken in the objects of the society was cordially endorsed, and the meeting then terminated.

NG 12 Sep 1879
The first show of NBKA opened on Friday in the Nottingham Arboretum, and continued on Saturday. The show is of a most interesting kind and it is somewhat significant of the condition of trade in the country that the principal reason advanced for bee culture is that it is highly profitable. A few years ago the idea of a small industry – as bee-keeping is, comparatively speaking – of an almost totally novel character being pursued systematically for the sale of the profits would not have been thought of seriously. Now, however, it is thought of with decided seriousness, on the principal that "every little helps." Still it is not at all certain that the general impression in prosperous times is the correct one; the American practice appears surely more practical, and if the apiary is to be made a source of profit it ought to be cultivated on the best principles, in order that the profit for which the cultivation is undertaken may be as large as possible.

This year especially, there is more than usual need for intelligence in the treatment of bees, since the insects are in a condition closely bordering on starvation where they are not fed artificially. To promote knowledge of the habits and instincts of the bee is the primary object of the Association, which is assuredly more worthy of patronage than many others which meet with the most ample support at the hands of the public.

The exhibition was held in the eastern wing of the Arboretum Pavilion, those operations which were necessarily conducted in the open air taking place in a paddock in the immediate neighbourhood, and also belonging to the Arboretum.

In the former place were shown various apiarian appliances of the most approved construction, for competition or otherwise, and also a large number of specimens of honey in the glass and comb and bees in exhibition cases.

The room was handsomely decorated lent by the Duke of St. Albans and Captain Rolleston, who also forwarded a number of specimens of honey. One of the most interesting objects shown in the room was a "honey slinger," by means of which the full comb can be removed from the hive, the honey removed, and the empty comb replaced, the advantage being the saving of one twentieth of their industry which the bees are calculated to spend upon the comb and in preparing it for the reception of the honey. Three or four different apparatus for feeding purposes were worthy of inspection, as well as some useful supers for taking the honey in the comb from the hive in quantities varying from one pound to ten.

The process known as "driving" from one hive to another took place every hour on Friday in the paddock.

The judges were Mr RR Godfrey, secretary of the Lincolnshire BKA and member of the BBKA and Mr Roberts of Belvoir, who acts for Mr Ingram, also of Belvoir; Mr WC Cheshire, the hon. secretary, rendering admirable service in all departments.

There was a good attendance of visitors during the afternoon and evening.

In the fourth class, a special commendation was given by the judges for a complete hive belonging to Mr Cheshire, and arranged on the cottager's principle but, as it was without floorboard, there was no prize awarded to it.
Bees
Class 1 – for the best stock of Ligurian or any other foreign bees – 1. F. Spencer, Beeston, Kramer bees
Class 2 – for the best stock of English bees – 1 and 2 WC Cheshire, Gedling
Special prizes for bees in any kind of hive, members and cottagers of Nottinghamshire – 1. David Woodhouse, Arboretum-street: 2. Simon Woodhouse, Mapperley
Special prize for best swarm – 1. TK Perrins, Mount House, Mount-street, New Basford: 2. WC Cheshire

Hives, etc

Class 7 – for the best honey extractor calculated to meet the needs of cottagers – 1. H Fuggle (only one exhibitor)

Honey, etc

Class 9 – for the best exhibition of honey in supers or sections of supers, separable, and each not more than two pounds in weight; the total weight of each entry to be less than ten pounds –2. Rev RB Garland, Bunbury, Retford

Class 12 – for the best exhibition of honey in the comb, taken from one hive without destroying the bees. Open to all *bona fide* cottagers residing in Nottinghamshire – no award.

Special prizes – 1. Mrs Musters, Annesley Hall: 2 and 3. WC Cheshire

Driving Competition

Class 13 – for the competitor who shall in the neatest, quickest and most complete manner drive the bees from a straw skep, capture and exhibit the queen; each competitor to provide his own bees – 1. Thomas Roberts, Retford; 2. WC Cheshire

The second day's exhibition of this society on Saturday in the Arboretum, was attended with most successful results, both as regard the influx of visitors and the interested manifested in it the manipulation of the bees and the working of the new and improved bee appliances.

The show of honey was a most creditable feature of the exhibition. Indeed, it was remarked by several beekeepers that the quantity placed upon the tables fell little short of that show at the exhibitions of much older societies. Some capital honey from America, Austria, Scotland and other countries was on view. In point of flavour the palm seemed to be accorded to the product of English hives.

Mr Godfrey, of Grantham, lent a charm to the exhibition in the shape of some beautiful illustrations of the bee in its various stages of growth, as well as the bee enemy. They were much admired, and attracted the attention of all interested in beekeeping.

A hive from Bavaria, constructed on different principle to those manufactured in England, drew around it a large number of admirers; but it was stated by some old beekeepers that the facilities for removing the bees were not equal to those made in this country.

Some new appliances for separating the honey from the comb elicited warm commendation. They are calculated to save much of the waste which has hitherto taken place in the old system. Some fine specimens of bees-wax were on view, indicating that great care had been taken in their production. A great proportion of the honey exhibited appeared to have found purchasers, and one splendid lot in 16 jars realised a high price.

During the afternoon and evening, at stated periods, the process of driving bees from one hive to another afforded much pleasure to those who had the courage to approach near enough to watch the manipulation. Mr Sells of Uffington, Mr Joseph Phipps of Arnold and some other beekeepers displayed some tact and coolness during the operation and, although at times literally covered with bees, escaped, we believe, without being stung. Several clergymen and gentlemen who take an interest in the society were present during this interesting part of the proceedings, and expressed themselves highly satisfied with the result.

Several of the committee, including Messrs. Phipps, Ward, Baldwin, Sissling and others, were most courteous in their attention to visitors and indefatigable in their in the endeavours to explain the nature of every exhibit to numerous enquirers. The secretary, Mr Cheshire, is deserving of praise for the successful result of the exhibition.

Altogether the promoters of the show may be congratulated upon the success, financially and otherwise, which has attended their first gathering, and the interest evinced augurs well for greater efforts and increased exhibits in future years. The committee and all who have taken an interest in the show are warm in their thanks to the Duke of St. Albans, Captain Rolleston, and other gentlemen who have aided them in bringing this, their first exhibition, to a most successful issue.

Driving is the art of compelling bees to leave their hive at the will of their master and is an operation often necessary with skeps to make artificial swarms and to clear honey-combs of bees when taking their stores. A typical setup is shown.

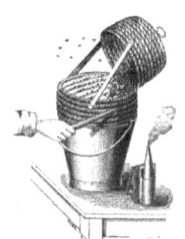

The bottom skep is given a little smoke and then is rapped with sticks heavily enough to jar the combs. The bees having gorged themselves with honey will move upwards in to the empty skep.

In competitions the winner is the first person to find the queen.

William Amelius Aubrey de Vere Beauclerk, 10th Duke of St. Albans (1840-1898). Bestwood Lodge was built in the Gothic style for the Tenth Duke of St Albans in 1862-65. Amongst his other titles he was Hereditary Grand Falconer of England.

John Chaworth-Musters paid for a school and a new church at Annesley in the 1870s, and Lina had a row of nine pairs of stone cottages known as 'The Grove' built on Derby Road in Annesley. He died from scarlet fever in France 1887. He was buried at Langar. His widow Lina lived

the remainder of her life at Wiverton. She was a writer and amateur historian, and a founder member of the Thoroton Society of Nottinghamshire.

Rev Thomas Bloom Garland, MA, Magdelene College, Cambridge was born in Leeds in 1835 and died at East Retford in 1908 aged 73. Curate of All Saints Church Babworth 1869 – 88. In 1888 he was at St. John the Evangelist's church in Bishopswood, Staffordshire with his sister, Mary. Lived latterly in Retford.

BBJ 1 Nov 1879

The first show of NBKA was held in the beautiful grounds of the Nottingham Arboretum on Friday and Saturday, Sept. 5th and 6th. The beekeepers of the 'Lace Metropolis' deserve great credit for their boldness in their endeavours to advance the knowledge of bee-culture, for we do not find many associations which in the first year of their existence have the courage to organize a two days' exhibition without any backing up from kindred associations. It was at first intended to hold the show in conjunction with one of the horticultural meetings of which Nottingham is by no means deficient, but the dates fixed for the horticultural displays were all judged to be too early in the season for a bee show; and so our Nottingham friends chose a bolder line of conduct and resolved to have an independent exhibition of their own; a resolve which was well supported by the kind manner in which Mr. Rogers, the lessee of the Arboretum Refreshment Room and Pavilion, offered the use of the latter for the purpose.

The pretty refreshment rooms of the Arboretum are fashioned in very humble imitation of the Crystal Palace, with a stone centre building having glass corridors stretching from it on either side some 200 feet or more and it was in the eastern one of these corridors that the hives, honey, and apiarian appliances, were exhibited, while the driving competition, and the manipulations generally, were conducted in a paddock adjoining, which on gala days is devoted to 'monster balloon ascents.' The Duke of St. Albans and Capt. Rolleston kindly assisted in the good work by furnishing the committee with splendid ferns and flowering plants for decorative purposes, and their horticultural beauties were freely and appropriately interspersed amongst the products of the busy bee.

In the classes for hives and bee furniture Messrs. Neighbour and Sons were in strong force, taking the first prize in Class 3 with their Philadelphia hives, and also first prize for the most complete collection of hives and bee furniture (Class 8.) Mr. R. Steele obtained second prize in Class 3 (for the best and most complete hive on the moveable comb principle) with his larger 'Eclectic Hive.' In Class 4, for the best complete hive on the moveable comb principle for cottagers' use, to include cover and floorboard facilities for storing surplus honey (price not to exceed 10s), the first prize was awarded to Mr. H Fuggle for a hive, which, if we mistake not, was marked at 6s 6d complete.

Straw hives on the moveable comb system by Messrs Neighbour, who exhibited a round hive with moveable slots, furnished with a super to match (somewhat after the Stewarton form), which, with floor-board complete, was marked at 4s 6d.

Mr. R Steele and Mr. H Fuggle both showed their extractors, the latter taking the first prize; for ' Extractors for Cottagers' use.'

Many other articles not for exhibition were shown, including sectional supers, slow and rapid feeders, feeding-stages, &c; of slow feeders the best was one of Neighbours, consisting of a tin can perforated at the bottom with five holes, which rested on a vulcanite stage in which a slot is cut to correspond with the row of holes in the can bottom. The can is fixed upon the stage so that it can only move backwards and forwards in the line of holes, and it is furnished with a pointer which moves over an indicator when the can is shifted, and shows on the indicator how many holes are over the slot in the stage; by a touch of the finger on the pointer the bee may be fed from one, two, three, four, or five holes at will.

In rapid feeders quite a large exhibit was shown by a local bee-keeper, Mr W Talbot, of Nottingham, one of this gentleman's exhibits (a bottom feeder) was made to fit in the floor-board, which it nearly covered, while a tube from the feeder was carried to the back of the hive and ended in a small covered funnel where the supplies could be poured in. Another noticeable article was a top feeder for the old-fashioned dome-topped skep: a hole was cut at the top of the dome, and a perforated zinc cylinder, about an inch deep, with perforated zinc bottom, is inserted, the cylinder being supported by a flange of zinc resting on the top of the dome, in this feeder a quart pickle bottle rested quite snugly and securely without any stage or other support whatever. When not in use for feeding the zinc cylinder could either be plugged or fitted with a ventilating but warm material.

In the classes devoted to honey exhibits, the show was very creditable, but the amount of this year's honey shown may be put down as nil, as the committee, finding that there were no entries made, very wisely withdrew the usual restriction in the interests of the show. Mr. Musters of Annesley Hall was awarded a special prize for a fine super weighing 16 lbs.

Amongst the honey exhibits, there was one of about a dozen 1-lb. jars of extracted honey shown by our friend, Mr. RR Godfrey of Grantham (who is always to the fore), but which was marked, 'Not for competition or sale.' This exhibit was staged by Mr. Godfrey in order to show what an attractive form our honey could he made to assume with a little care. The jars were of an inexpensive but clean and pretty character, with nice glass lids, each jar being tastefully finished off with a neat label. The exhibit was a capital lecture in itself, and one which it is to be hoped the Nottingham beekeepers will profit by. The writer repeatedly heard the question asked if these jars of honey were for sale; and they could no doubt have been sold over and over again, simply from their attractive appearance.

Samples of American honey, both in sections and in jars, were also shown by Mr. Godfrey, who kindly allowed many of the visitors to taste the difference in flavour between the imported article and that obtained from English clover or heather. Mr. Godfrey's diagrams, showing the anatomy and physiological characteristics of the honey-bee and kindred insects, occupied a conspicuous place at one end of the large corridor, and claimed a large share of the attention of visitors.

Proceeding to the paddock where the driving, &c. took place, we found ourselves in a small square field of about a quarter of an acre, round two sides of which were about thirty hives of all kinds from Abbott's Standard down to the antediluvian skeps. Neighbour's Philadelphia and their well-known No. 5, Pettigrew's big skeps, and home-made hives of various kinds, were here shown full of bees, some for competition, some for driving purposes.

There was one novelty here noticeable in the shape of a German moveable comb hive containing bees, brought from the Austrian Tyrol, and which the owner, Mr. F. Spencer, of Beeston, called 'Khraner bees.' The hive itself was an object of interest to the beekeepers present, and caused some free comparisons to be made between it and our own bar-framers. The hive, which stood upon a four-legged stand, had very much the appearance of a gas meter, or rather the box with which our household meter is covered. When the door of the hive was unlocked and opened (in a manner which bore out the meter resemblance very much), the whole of the interior was seen to be like a small cupboard with one shelf dividing the lower and longer portion, which is used as the brood nest or pavilion, from the smaller upper portion, which is the supering apartment. Both these divisions are fitted with frames, which slide in from the front, and rest on rabbets, top and bottom. Thus the top of the hive is never removed, and the frames cannot be lifted out as in the English bar-frame hive, but are always fished out from the front with a pair of pliers.

The first prize for foreign bees was awarded to Mr Spencer for the bees in this hive. For English bees in observatory hive, Mr WC Cheshire, the hon. sec. of the Association, took first and second prizes.

The driving competition began about 2pm, (by which time a considerable number of visitors were present), and was continued all the afternoon; the successful competitors being Mr. Thos Roberts, of Retford, who drove and captured the queen in eight minutes, the second prize was awarded to Mr WC Cheshire, of Gedling.

BEEKEEPING IN VICTORIAN NOTTINGHAMSHIRE

Most of the bees which were driven during the day were transferred to bar-frame hives by Mr J Phipps, of Arnold, and other members of the Association, who, with their shirt-sleeves rolled up, astonished the visitors by the facility with which they cut up the hive and transferred the combs and living bees. On the second day, Saturday, a number of hives were driven and transferred for persons who had purchased bar-frame hives at the exhibition, and had brought their bees to be put into them.

The driving operations were, as usual, the most popular part of the exhibition, though it required some courage on the part of the strangers to approach the operators, for the only protection provided was a gauze fence about eight feet high over which the bees of course came and scoured about the fields by thousands. If the perfect security of the BBKA's bee tent had been afforded, the enjoyment of this, the most instructive part of every bee show, would have been complete.

Notwithstanding this slight drawback the general arrangements of the exhibition were very good, and reflect great credit upon the committee of the Association and its secretary, Mr. Cheshire, to whose indefatigable exertions the success of the exhibition was mainly due. The members of the committee, including Messrs. Phipps, Ward, Baldwin, Talbot, Sissling, and others, were most courteous in their attention to visitors, and were busily engaged both days in explaining the nature of the exhibits to numerous inquirers.

The sales effected were very satisfactory, the exhibits of Messrs. Fuggle and Steele being all cleared off, and several of Messrs. Neighbour followed suit. The whole of the honey staged for sale was disposed of and the cry of visitors was still for more.

Although it is to be feared that the committee will be losers financially by their praiseworthy effort to advance the cause of humane bee-culture in this district still the good seed has been sown, and the readers of the BBJ may safely consider Nottinghamshire as another district in which the humane culture of the honey bee has become and established fact.

The judges were Mr RR Godfrey of Grantham and Mr T Roberts of Belvoir, acting for Mr W Ingram, who found himself unable to attend.

The Bienen Zeitung explains: "Krainer" bees. The rough climate of our mountainous country, Austria, has made our bees a hardy race, for they have been hemmed in by mountains for centuries. Our bees fly in dark and cool weather, and suffer very little from diarrhoea, and other spring maladies The rich mountain meadows and forests of Krain, with their profusion of bloom, animate our bees in the spring, and we often find them flourishing in March and April, breeding early, and beginning early to swarm. Our bees are proverbially kind and gentle. These bees are great beauties, being nearly black, with white stripes. The white being probably due to hairs, as they all seem to be young bees.

BBJ 1 Dec 1879
Nottingham. There are no echoes from our hives here; all are in profound repose, being robed in a white mantle of snow - so far, however, all is well. I have not heard of any losses yet, although where hives have not been fed the present outlook is very poor indeed. WS.

BBJ 1 Jan 1880
Nottingham, Dec. 20, 1879. Jack Frost has been the ruling power nearly all the month. December 1st brought a heavy snow-storm, which continued on the 2nd, and left us with a ground covering of three or four inches of snow. On the 6th the thermometer was several degrees below zero, showing the lowest temperature since 1860 and so frost and fog have ruled all the month except the 21st, 22nd and 23rd. On the 22nd we had a beautiful spring-like day, mild as April, which made the birds sing and brought out the bees in hundreds. This gave beekeepers an opportunity of overhauling their stocks.

BBJ 30 Jan 1880
The Annual General Meeting of NBKA will be held on Saturday next, Feb. 1st, at the People's Hall, Heathcote-street, Nottingham. The chair will be taken by Viscount St. Vincent at 3.30pm. A knife-and-fork tea will be provided at 4.30 pm. Tickets to be obtained of the Honorary Secretary, Mr. AG Pugh, Mona-street, Beeston. A social evening and prize-drawing are to follow.

Arthur George Pugh was born in Tuffley, Gloucestershire in 1856. In the 1881 Census he was employed as a Permanent Wagon Sub Inspector. By 1891 he had become a Permanent Way Inspector living with son, Thomas Arthur, who was born in Ambergate in 1889. In 1911 he was a contractor with his son Thomas A, helping him.

Sadly, during research for this book in 2018, it was found that Mona-street has been almost completely demolished including Arthur's house, and replaced with a housing estate.

BBJ 1 Feb 1880
In presenting their Annual Report for the year 1879 to the members of the BBKA is was said that there are now eight county Associations affiliated with the Central Society, viz. those of Dorsetshire, Devonshire, Hertfordshire, Lincolnshire, Shropshire, Nottinghamshire, Surrey, and West Kent. These County Associations, by the payment of one guinea annually, are entitled to:
(1) The free use of the Bee Tent at their Annual County Shows, and to all the proceeds accruing from the same.

(2) To a Silver Medal, a Bronze Medal, and a Certificate of the Association, to be offered as prizes for honey, more especially for the production of honey in the comb. These medals and certificates have been offered for competition, and awarded at the County Show held in Nottinghamshire.

The Secretaries of the County Associations are requested to furnish the Honorary Secretary with the names of the winners of these medals and certificates in future years, with the view of their being included in the yearly Report.

NEP3 Feb 1880
A committee meeting of NBKA was held at the Mechanics' Hall on Saturday. The Rev TW Smith was in the chair in the absence of the president. Messrs Phipps, Cheshire. Sissling, Woodhouse, Ward, Baldwin, Talbot, etc. were also present. The minutes of the last meeting being read over, and the accounts of the past year being examined, the secretary thought it advisable to have lectures delivered in the different districts around Nottingham, and a final one or more in the Mechanics' Institute to encourage persons in the humane principle of keeping bees, and to save their present stock from perishing.

BBJ 1 Mar 1880
Nottingham, Feb. 23rd. On January 31st, we had the first indication here of returning spring, the day was bright and sunny, and the bees made their appearance for the first time this year. February, so far, has been mild and showery, with bright gleams of sunshine the middle of the day. Our bees are now out daily; but there are, of course, no flowers yet for them to visit. They, however, accept pea-flour when offered to them in bright weather, and all seem to be going on well. WS.

BBJ 1 Mar 1880
Queen-raising. I wish your advice in regard to raising queens, and keeping them for future use. I have for some years practised artificial swarming, and have often had four, five, and six spare young queens; and what I want to know is, how can I keep them after they are hatched out? Say the hive is examined a few days after artificial swarming, and contains five royal cells, one will be required for the parent hive; and if my other hives are not ready for swarming, or a wet time sets in, and I am prevented from cutting out the four other royal cells, how shall I prevent the worker bees or the first queen hatched from destroying them? What is usually done in such a case? Is it the proper wav to place a small wire cage over the royal cells pressing it into the comb and if I did so would the worker bees feed them while so engaged if they had a young queen at liberty on the combs? Again, would they be likely to eat away the comb, and let them out? Or is the usual way to encage them on the top of the hive? If so, would you describe the kind of cage or box in which they are placed, and if the royal cells are put in before they are hatched out; or in other words, in what stage of

development are they put in; and will the worker bees feed them so encaged?

G. Green, Eagle, near Newark.

Reply:
The obvious mode of preventing most of the ills prospected is to delay the first artificial swarming until the other stocks are within a few days of being fit for similar treatment. Natural swarming, however, often occurs, and places one in the difficulty indicated. Ripe, ie. fully sealed, queen-cells should be removed by cutting out the portion of comb upon which they are raised, if it is intended to preserve them. They should then be fastened to a piece of comb with honey on it, and placed in a box with a handful of bees, and kept in a warm place with a moist atmosphere. No better place can well be found than the top of their own hive, where, provided the heat of the hive passes through the box, the hatching will duly take place. Four boxes, say three inches square (cube), with a wired corner hole, might be so arranged with the corners together over the ordinary feed-hole, that each would get a share of the heat from below. It is not safe to trust to worker bees feeding unfertile queens, they will generally feed themselves. It is better to catch the young queens as they hatch, if possible, and imprison them, and so prevent cutting the comb. Ed.

BBJ 1 May 1880
Nottingham, March 29, 1880. March has up to this day been a splendid month here. We have bright warm sunshine every day, though the nights are still chilly. March 4th was the first day our bees were noticed bearing pollen, and since then they have been at work every day. The purple crocus is now in full bloom in the fields and the spring flowers and flowering shrubs in the garden afford plenty of pollen and possibly some honey during the sunny mid-day hours. The season promises to be very favourable so far. WS.

Southwell, April 20. I have wintered fourteen hives out of eighteen, having thus lost four, one of them being queenless. I am the only one in the parish of Southwell that has any bees left: they have all paid the penalty of letting them alone. I have given my bees at least 2 cwt. of sugar during last autumn and this spring but I am well repaid for my trouble and expense. My bees have lived entirely on sugar syrup during the winter, as most of my hives were entirely honeyless; so bad was last season, that had I not fed liberally I should not have had one stock left. I trust the coming season will cheer us all up. SS.

BBJ 1 May 1880
Retford. It is, I believe, usually asserted that bees gather only one description of pollen on the same journey. Today I noticed a bee returning to her hive having on each thigh two balls of pollen of different colours. Is not this very uncommon?

TBG (Rev. TB Garland?)

BEEKEEPING IN VICTORIAN NOTTINGHAMSHIRE

BBJ 1 Jun 1880
Nottingham. Drones were first seen on May 20th, so that no swarms have issued yet, although the weather during the present month has been all that could be wished. April, however, was a cold month, and the progress made was not as rapid as we had been led to hope from the state of our stocks in March. All at present, however, seems to presage a good season both for bees and honey. WS.

BBJ 1 Aug 1880
Nottingham, July 24th, 1880. Some of our friends in your July issue speak of taking big supers, filled with honey in a most tempting manner; but I fear that our beelines have not fallen in such pleasant places. We have plenty of bees, but the weight of honey sealed up to this date has been very small. June was a wet month, and July has been a very stormy one, every fine day being usually followed by three or four wet or indifferent ones, so that I am afraid we shall not have any brilliant results to show for our pains in 1880. My first swarm issued on May 27th, the last on July 13th. This was from a skep which I had doubled on Mr. Pagden's plan on July 1st; but instead of building comb in the lower hive as I expected, they clustered in it for thirteen days and then swarmed suddenly. I gather from this, that nadirs are not certain to prevent swarming any more than supers. On June 27th I transferred a stock to a bar-frame hive (this was twenty-four days after swarming), and the weight of honey taken from the transferred comb was 7 lbs. Killing drones commenced on July 10th, and still continues. WS.

BBJ 1 Mar 1881
BBKA Annual Report. It is to be regretted that no communication has been received during the past year from the Nottinghamshire County Association, formed in 1879, and that the Affiliation Fee has been suffered to lapse. It is hoped that the bee-keepers and others interested in the promotion of bee-culture in the County of Nottingham will exert every effort to revive their County Association, and establish it on a firm basis.

BBJ 1 Nov 1881
Hon Sec BBKA. I may add that the Lincolnshire Secretary, Mr. RR Godfrey, writes to me that he is inundated with communications from beekeepers in the counties of Lancashire, Leicestershire, Northamptonshire, Rutland, Nottinghamshire, Huntingdon, and Cambridgeshire, who ask for advice and information in endless ways. 'I would suggest,' he says, 'that the BBKA should storm each county during the coming winter by means of meetings held on a bold scale, which should be held on a market day in the various market towns in each county, 'The BBKA would do a grand stroke,' he says, 'if it could bring about the formation of Associations in the counties named.'

NG 24 Mar 1882

The Apiary. The bee-keeper must now be on the alert and carefully ascertain the condition of each hive, and take such steps as will place the bees in the best possible condition for taking advantage of every opportunity for gathering honey. It is most important that stocks short of food should at the present time be fed somewhat liberally. The bees are now becoming active, and naturally require more food than they did a few weeks back, and as yet their feeding ground is somewhat limited. Artificial feeding is not desirable for augmenting the supply of honey in the hive — if a mixture of sugar and water after being stored by the bees can be so designated — and it ought not to be resorted to for that purpose. But the supply of artificial food for the purpose of maintaining the strength of the stocks is perfectly legitimate, and can be strongly recommended. It is important some time during the month of March that the hives should be examined for the purpose of ascertaining their condition; and in the case of stocks that are weak, an early opportunity should be taken advantage of for uniting two together, as one really strong stock is very much better than two weakly ones, as will be seen later on in the season.

BBJ 1 Jan 1883
A most interesting experiment has been inaugurated by the Hon. and Rev. H. Bligh, which has for its object a comparison of the different hives and the most economic modes of working them. Seventeen competitors have entered the lists, and the matter will be settled at the end of August in the present year.

BBJ 1 Jan 1883
GW Coxon, Nottingham. We were very pleased to receive your communication; but your kind suggestion has been anticipated, another subscriber having given its instructions that the bee-keeper who was obliged to give up the Journal through failure of work should be supplied with it till the arrival of better times.

George William Coxon was a Labourer in a Mechanics Shop in the 1881 Census. He was born in 1862 in Nottingham. In 1907 he was living in Southwell

Peter Scattergood. In 1880 was a machine builder and churchwarden. He was reported in 1882/3 as Clerk to Stapleford School Board. In 1886 he had for sale a patented warp machine from Whitehall's factory. He died in 1908 at Gladwin House, Stapleford. His many activities in beekeeping both locally and nationally appear throughout this narrative.

BBJ 15 Jun 1883
Death from a Bee-sting. On Monday evening, the 11th inst. an inquest was held at Torworth, near Retford, on the body of a farmer. It was shown in evidence that the deceased, when in his garden, was stung on the forehead by a bee, and that he died almost immediately from the effects of it.
BBJ 1 Jul 1883

BEEKEEPING IN VICTORIAN NOTTINGHAMSHIRE

Burrowing Bees.
Will you kindly tell me in your next issue what kind of bees the enclosed are? They have appeared in large numbers this spring in the garden of a gentleman of my acquaintance, and look so much like the hive-bees, when carrying pollen, that I at first thought they were so. They have burrowed under a lawn close to the house door, entering at the edge of the walk. Some have made their home under the turf, while others have turned after going downwards, and burrowed under the asphalte walk, where we found cells with pollen in, at a depth of from twelve to fourteen inches. I enclose several cells with the pollen rolled into round pellets just as we found them.
William Coxon, Nottingham.

[We are obliged to our correspondent for the trouble he has taken in forwarding the cells of the burrowing bee. They are very perfect specimens of 'insect architecture.' They are composed of sand-grains, glutinated together by means of some viscid saliva; thimble-like, rough on the outside, but perfectly smooth in the interior; with pellets of mingled pollen and honey about the size of a red currant, the instinct of the bee teaching it the amount of nurture required for the subsistence of the young which shall proceed from the egg to be deposited. The species is the *Andrena trimmerana*, it belongs to the family of the *Andrenidae*, of which there are upwards of eighty native species in Great Britain.]

BBJ 1 Aug 1883
Nottinghamshire Association. The Rev. AH Halley, of Cotgrave, Notts, has undertaken to receive the names of persons who are desirous of becoming members. Steps are now being taken to establish an Association for Notts on a sound and firm basis, and we trust all those who were members of the former Notts Association will send in their names to Mr. Halley without delay.

Alexander Hay Halley, vicar of Brockham 1901-08, was born about 1858. He entered the Canons' Theological School, Lincoln in 1878; deacon 1880, priest 1881. He was curate of Holy Trinity, Louth 1880-82, of Cotgrave, Nottinghamshire, 1882-84 and died in 1922.

NEP 7 Sep 1883
It is proposed to form a Beekeepers' Association for the county of Nottinghamshire, in connection with the BBKA. The object of the association to encourage and advance bee-keeping amongst the residents of the county, more especially amongst the agricultural and other labouring classes, and afford to members as to the most profitable manner of managing their bees, and disposing of their produce. In a favourable season a cottager may make a considerable addition to his income by keeping bees, if he knows how to manage them, and if he can be certain of finding a market for his honey. Those wishing to join the proposed association may apply to the hon. secretary, the Rev. AH Halley. Cotgrave, Notts.

BBJ 15 Sep 1883
Nottinghamshire Bee-Keepers' Association. All who would like to join this Association (in connexion with the parent Association) should send their names to the Rev. AH Halley, Cotgrave, Notts, who has promised to act as Secretary until the Association is fully formed.

BBJ 15 Sep 1883
Rev. AH Halley, Notts. Best Sugar for Bee-food. That sugar is the best for bees which has the largest amount of saccharine matter and is least chemically dealt with. We therefore consider the sugar most to be preferred is the old-fashioned Muscovado raw sugar—the Porto Rico or Dry Barbadoes. Duncan's Pearl (a fine dry sugar) and Refined Dry Crystals may also safely be recommended. Unrefined sugar is, however, more subject to fermentation than refined. In loaf sugar (refined from raw sugar), and in Demerara, some colouring matter is introduced—a blue into the former, and sulphuric acid into the latter.

BBJ 1 Oct 1883 Adverts
Seven Stocks for Sale in Frame Hives, 8 frames each, 16 by 8, all full of Comb; Hives well painted, with outer loose Covers. Also 2 Empty Hives, same size; and 2 Stocks in Straw Skeps. 1 Empty Skep. New Smoker, Veil, and sundry spare Frames, Covers, etc. £9 the lot, or separate.   J.Ward, 138 Woodborough Road, Nottingham

Several Stocks, English and Hybrids, in double-walled Hives, cork-dust between. 12 Frames and Dummy. Sufficient food for winter. On rail, 23s each. Pure Ligurian, 30s.
        A. Simpson, Mansfield, Woodhouse, Notts

BBJ 1 Nov 1883
A Hint to Nottingham Bee-keepers. The Rev. TF Boultbee, late of Escot, Ottery St. Mary, an enthusiastic and successful bee-keeper, and a valued member of the BBKA, has been presented to a living in Nottingham.

Thomas Francis Boultbee (1850-1925) married Marian Gertrude Padwick (b. 1844 Thorney, Sussex) in 1876.

NG 28 Sep 1883
St. Mark's, Nottingham. The Vicarage of St. Mark's, Nottingham, vacant by the resignation of the Rev. W. Feiton, has been accepted by the Rev TF Boultbee, vicar of Larkbere, Ottery St. Mary's, Devon.

NG 10 Oct 1883
Sir John Kennaway presided, and after expressing the regret universally felt at his departure, and the earnest wishes and prayers that would accompany Mr. Boultbee in entering upon his

responsible work in a parish of more than 8,000 souls in Nottingham, presented, in the name of 150 subscribers, a handsome silver tea-kettle, with the following inscription " Presented to the Rev. TF. Boultbee, MA, Vicar of Escot, 1877-83, by his parishioners and friends, in token of their affectionate regard and their value for his ministry." A number of handsomely-bound books were also presented. Mr. Boultbee responded in warm terms.

BBJ 15 Nov 1883
WH Radford, Nottingham. Flour-cake. Pea-flour cake at this time of year is a mistake, as the consumption of nitrogenous food requires cleansing flights, which in chilly weather leads to loss of life. Remove it and give it plain cake for the present, say until middle of February, if open weather. The sample you send is a little burnt, but not so much so as to be injurious. It will keep through the winter, if stored in a dry place. You say you put the cake over the quilt; you should put it under it, between it and the frames.

T. Rose, Radcliffe-on-Trent. To convert round-top hives into flat ones it would be necessary to cut off a few inches, and sew the raw edge with string. Then proceed to adapt it to the super-case as directed in the Association's tract 'Skeps' (price 1d). The cutting off the top need not occupy above five minutes, and a puff or so of smoke would render the bees quiet during the operation. The bees would forthwith refix the disturbed combs. This might be done in the spring.

BBJ 1 Dec 1883
Viscount Newark has consented to be the President of the NBKA.

Charles William Sidney Pierrepont, (1834-1926) 4th Earl Manvers. He served as the first president of NBKA from 30th April 1884 to 1st February 1890. During this period he had not yet succeeded to the earldom and he was styled Lord Newark, or more correctly, Viscount Newark. He remained a vice-President till at least 1901, paying his generous subscription of £10 pa as a member living in Holme Pierrepon Hall. The local village was anmed after him. He moved to Thoresby Hall in 1900.

A Meeting of NBKA was held in the Exchange Buildings, Nottingham on 30th April, 1884. In the absence of Lord Newark, who had promised to be President of the Association, the chair was occupied by the Rev HP Ling; the Rev HR Peel and Mr J Huckle, Secretary of the BBKA, attended the meeting on behalf of the parent Society.

The Rev AH Halley reported that he had corresponded with upwards of eighty residents of the county, most of whom had promised to support the Association. The following resolution, moved by the Rev HR Peel, and seconded by Mr. Beeson, was carried unanimously, viz. "That it is desirable that a Beekeepers' Association be established for the county of Nottingham, and be affiliated with the BBKA." The Rev AH Halley was unanimously elected Hon. Sec. *pro tem*, and a representative committee was formed, including several ladies. It was resolved that meetings be held in various parts of the county, and the Central Society be requested to send a lecturer to address these meetings.

There are two John Beesons in the 1881 Census for the Ratcliffe-On-Trent area – one born in 1830 who was a boat owner and shopkeeper and the second born in 1845 who was a general shopkeeper.

One would have been 54 years old at the time of this meeting and the other 39 years old so it is not clear which one attended.

NG 1 May 1884
Beekeepers' Association for Notts. Yesterday afternoon a meeting was held in the Exchange Hall, Nottingham, for purpose of forming a Beekeepers' Association for Nottinghamshire. There was moderate attendance, and the Rev. TP Ling presided. There were also present Rev. R. Peel, secretary of the BBKA; Rev. G, Shipton, of Chesterfield; Rev. AH Halley, Cotgrave, hon. secretary, and others. Rev. Halley stated that they had been in correspondence with 80 people in Nottinghamshire, and most of them were willing to become members of the society. One of their great objects should be to get the Agricultural Society to help them, giving a subscription and setting a space their disposal at their annual shows. They were grateful to Lord Newark, who had promised to be their president. The Rev. R. Peel moved the following resolution: "That it desirable that a Beekeepers' Association should be established connection with the County of Nottinghamshire, and that be affiliated with the BBKA." At the county shows they offered silver and bronze medals and certificates to be competed for by those who exhibited honey, especially honey in the comb. They also offered them the free use of one of their bee tents. They desired, as far possible, to encourage cottagers, and give them the means of adding to their scanty incomes, while they also supplied them with an innocent, harmless pursuit. Another object was to raise up national industry. Tons of honey were imported into this country from Switzerland, Germany, Italy, France, and America, and they would find that the greater portion of the honey that was imported was not pure honey, which was good and beneficial for man, very often was not honey at all, but consisted of grape sugar or corn sugar, or some compound that had nothing to do with honey. They wanted to enable their fellow countrymen to create national industry, and produce for themselves what they at present imported in a vitiated and adulterated form. (Hear, hear,) A further object of the association would be to prevent cruelty to the bee, and the Society for the Prevention of Cruelty to Animals helped them freely. Mr. Beeson seconded the resolution, which was unanimously adopted.

BEEKEEPING IN VICTORIAN NOTTINGHAMSHIRE

The Rev. G Shipton described in detail the process of honey making and the habits of the bees, and expressed hope that they might be the means of extending to others the advantage they found in this innocent and profitable source of employment, especially to working incidentally mentioned that sum of £150,000 was sent out of this country last year for honey and wax, and if they could bring that money into the pockets of agricultural labourers and others great result would have been attained.

The Rev. AH Halley explained, reference to over stocking, that the dearth of spring flowers in early spring had led people to supply that want by sowing seeds of spring flowers in great abundance.

The following officers were elected: President. Lord Newark; treasurers, Messrs. Smith and Son, bankers; secretary, the Rev. AH Halley: Committee: Revs. AJL Dobbin, HP Ling, and PF Boultbee, and Messrs J. Mann, A. Lewis, J. Barron, and Sandaver. A vote of thanks to the Chairman brought the meeting to a close.

Rev Abraham Joseph Locke Dobbin, BA, was vicar of Cropwell Bishop from 1877. Prior to this he was curate of Ruddington. He was born about 1836 and married in Oldham in 1864 and had three children. He died in September 1900 at Bingham aged 64.

When, on 1st June 1897, the Thoroton Society held its founding meeting at the Shire Hall the Duke of St Albans became the President. A council of twelve members was elected, with a chairman – Lord Hawkesbury – who never attended another meeting! In his place the chair was usually taken by the Rev Dobbin. He was a member of the Cropwell Conservative Association as was Mr Mann. He was also a Freemason (1885). He was organiser of candidates for the Board of Education for diocese of Notts in 1887.

Rev Henry Pratt Ling MA was rector of Keyworth, St. Mary Magdelene (1878-1928) and Patron there 1873 – 1930. He was also rector of All Saint's, Stanton-in-the-Wolds 1892 – 1928. He was born in 1849 in Braintree. He disappeared from Crockford's Register of the Clergy in 1936.

BBJ 15 May 1884
Bligh Competition, 1884-5.
This interesting competition commences on the 20th inst. We are pleased to report that the entries made are double in number compared with the previous contest. The following counties are represented, viz., Herts, Cornwall, Kent, Cambridgeshire, Susses, Berks, Worcestershire, Wilts, Notts, Derbyshire, Bucks, Essex, Surrey, Herefordshire, Oxfordshire, Lincolnshire, Middlesex, and Gloucestershire. The entries made are principally on the frame-hive system. There are, however, several on the skep system, and the Stewarton is also represented. Each competitor should note that full details of the mode in which he commences the competition should be entered in his diary.

> Do not order your Hives and Appliances until
> You have seen EC Walton's CATALOGUE
> For 1884, consisting of 36 pages, fully illustrated.
> Sent free for 1d stamp on receipt of address.
> EC WALTON
> NORTH MUSKHAM, NEWARK
> Advertisement BBJ 15th May 1884

BBJ 15 July 1884
Bath and West of England Show – EC Walton was listed as an exhibitor.

BBJ 15 Aug 1884
Lincolnshire BKA Show. Best 12 1-lb sections of comb honey in crate – Rev Chas Plumtre, Claypole, Newark.

The restored Plumtre Hospital established by this family is situated on London Road, Nottingham

BBJ 15 Sep 1884
Could you give me any information about the Nottinghamshire Beekeepers' Association? I was at the first meeting when they formed it, and paid my subscription, but have not heard anything more about it, except that the Hon. Sec. (the Rev AH Halley) has left the county. I hope it is not falling through, as I think it would be a great boon to the Notts. beekeepers. I keep looking very intently for the advertisement about the annual show and the bee-tent engagements for Notts in every issue of the British Bee Journal, but have seen none yet. I hope I may see one before long.
Thomas Rose, Radcliffe-on-Trent, Nottinghamshire
(Perhaps the insertion of the above may elicit some information as to the present position of the Nottinghamshire BKA.)

The only Thomas Rose I could find was born in Radcliffe-on-Trent in 1863. He was the manager of the Co-operative store in Netherfield. He may have married Sarah Hemsley in September 1886.

BBJ 15 Dec 1884
Bligh Competition, 1884-5.
Now that the first of the two seasons over which this Competition extends is fully past, it may be well that we should report progress. The entries were far more numerous than on the former occasion, and amongst the thirty three was registered, the Rev TB Garland's, Retford, frame was one of them.

Of these, however, we regret to say that no fewer than fourteen have already withdrawn from the Competition. Of the remaining nineteen, only eleven have sent in their two-monthly reports up to the end of October.

NEP 19 Dec 1884

The annual meeting of NBKA was held on Thursday afternoon at the Exchange Hall, Nottingham. Alderman John Manning JP presided, and among those present were: The Rev R Holden, Rev TF Boultbee, Rev TB Garland, Messrs J Brierley, Price, Rose, Beeson, RR Godfrey, Mrs Wotton, Mr Ferneyhough (hon. Secretary), &c. The Association was established in the year 1884. Its objects are to encourage amongst the residents of the county, and especially the cottagers and labouring classes, a more humane, intelligent, and profitable system of bee-keeping. The Association is a branch of the BBKA (or Central Society).

The annual report of the society states that Nottinghamshire is a very good honey yielding district, but a large quantity of honey is annually wasted owing to a want of proper knowledge of bee culture. The cottager's honey is, from the manner in which it is taken, often almost unsaleable, and has hitherto been driven out of the market by foreign honey. English honey will always command a better price than foreign, much of which is adulterated and is entirely devoid of the fine aroma and flavour of pure English honey. Good honey, put up in neat and attractive form, commands a ready sale and realises good prices.

Among the rules of the society are the following:
"District secretaries shall be empowered to select, or cause to be selected, one or more representatives of each parish in their district, and shall summon them to attend meetings, either monthly or as often as may be found convenient. At these meetings, after business affecting the Association has been disposed of, papers may be read, and questions connected with bee-keeping may be discussed."

"The parish representatives shall furnish the district secretary with a list of all the bee-keepers in his parish, and shall induce as many as possible to become members of the association. They shall make any suggestions with regard to the working of the association, which they may have to offer, to the district secretary."

The first business before the meeting was the election of officers for the ensuing year as follows:
President, Viscount Newark; Vice-presidents, Ald. Manning and Mrs Robertson (Widmerpool); hon. secretary, Mr E Ferneyhough; hon. Treasurer, Mr HE Thornton; committee, Mr J Barron, Mr J Beeson, Rev HJL Dobbin, Mr J Geeson, Mr A Felstead, Mrs J Wotton, Mrs Hole, Mrs Mason, Rev AH Halley, Mr A Lewis, Rev HP Ling, Mr W Silver, and Mr J Mann.

It was proposed, on the proposition of the Chairman, that the committee be authorised to appoint district secretaries in connection with the association. The Chairman also observed that although he was not himself practically acquainted with bee-keeping, yet he thought that valuable services might be rendered towards the promotion of bee culture by the district secretaries, and especially by the clergy of the various parishes.

Mr RR Godfrey of Grantham (Hon. Sec of Lincolnshire BKA), gave an address, in which he expressed the great pleasure which he felt in attending the meeting. He expressed an opinion that each of the district secretaries should be acquainted with the culture and management of bees, and that they should be responsible for the different local exhibitions. He felt sure that Nottinghamshire was one of the most suitable counties in which the operations of the Bee-keepers Association could be carried on. He spoke of the great success which had attended the holding of a honey fair at Grantham, and he recommended that each county should give its own prizes for the best honey. (Hear, hear!)

A short discussion ensued, and a vote of thanks was passed to Mr Godfrey, and to the chairman for presiding.

The only Ferneyhough I have been able to find in the local records is Ebenezer – born in 1850 in Nottingham, married in June 1873 and died in 1933 aged 84.

Rev Robert Holden MA (opposite) was born in Spondon in 1853. He was Rector of Nuthall and held many positions in the Southwell diocese.

William Silver, Bridge-gate, Retford. He was born in 1854 in Reading, Berkshire and is employed as a coach builder manager at this time. He qualified as a BBKA "expert" in 1888.

Mrs Harriett Annie Robertson born 1839 and died in 1891 aged 42. Wife of Major George Coke Robertson of Widmerpool Hall. Her memorial in St. Peters Church is extremely lifelike. The spire at the hall does not have a clock in respect of her memory.

Mrs Caroline Franklin Hole born 1841 and married Rev Samuel Reynolds Hole, vicar of Caunton, in Chelsea in 1861. She died in 1916 at Hawkshurst, Kent.

BBJ 1 Jan 1885
Tits and Fly-catchers. The Rev. CS Millard, Costock Rectory writes:
I am sorry to see that your correspondent, HF Hills, advocates shooting flycatchers, some of the most useful and interesting birds we have. Is he quite sure that he has not mistaken friends for foes? My own experience would lead me to think so. We

have many fly-catchers here, but only once or perhaps twice have I seen them paying attention to my bees.

On one occasion I was much concerned to see one perched on one of my hives, then darting every minute or so to the door of the hive, and carrying off one of my bees to feed a young one which sat on a high wall at the back of the hives. I neither shot nor threw stones at them, but watched. Presently the young bird flew to the ground with the bee its mother had just given it, and I ran up before it had time to swallow a fat drone that was struggling on the ground. If it was only drones they were killing they were quite welcome to them, and I am inclined to think this was the case as the old bird seemed to single out her prey. A drone would be a much safer mouthful than a worker-bee. I have never seen tits at my beehive, but one of my parishioners tells me he has watched them tapping at the hive door and killing the bee that answered the knock.

NG 10 Apr 1885
Lecture at Hucknall Torkard. On Tuesday a lecture was delivered on "Beekeeping," by Mr W Bell, of Annesley, in the Assembly Boom at the Coffee Tavern. The chair was occupied by the vicar (the Rev. JE Phillips), and the lecture was listened to with great interest by the audience, among whom were a number of local beekeepers. The lecture was delivered under the auspices of NBKA

BBJ 15 Apr 1885
Nottinghamshire. Rev. AH Halley. The number of members of this Association is now 71.

BBJ 15 Apr 1885
A lecture in connexion with NBKA was delivered at the Board Schoolroom, Willoughby-on-the-Wolds, on Tuesday, April 7th, by Mr. Frank HK Fisher (Farnsfield), on 'Bee-keeping: How to make it Pay.' In the course of a very interesting lecture, Mr. Fisher explained the many advantages of the modern bar-frame over the old straw skeps, and showed how bees may be kept to produce a large profit. Major Robertson, of Widmerpool Hall, occupied the chair, and kindly offered a prize in connexion with the coming show at Willoughby. There was a large and appreciative audience. A vote of thanks to the lecturer and the chairman brought the proceedings to a close.

A lecture was delivered on Tuesday evening, April 7th, at the Coffee Tavern, Hucknall Torkard, by Mr W Bell, of Annesley, on 'Bees and Bee-keeping: How to Make it Pay.' The Rev. JE Philips, vicar of Hucknall, occupied the chair. The lecturer thoroughly explained the working of a bar-frame hive, and showed its advantages over the old straw skep. In the course of his remarks Mr. Bell enumerated the many benefits to be derived from joining the Association, which already numbered over seventy members in various parts of the county. There was a large audience, who were deeply interested, especially with the practical illustrations introduced by the lecturer. A vote of thanks to the lecturer and the chairman brought the proceedings to a close.

The Rev. John Edward Philips, MA, came to Hucknall as curate in 1875, and was created vicar and patron in 1879. On the resignation of Canon Pavey, Mr. Phillips was created Rural Dean of Mansfield, in compliance with the expressed wish of the clergy. To Mr. Philips' memory the parishioners and friends placed a brass and pulpit in the church, and a granite cross on his grave. The Revs. JB Hyde, JE Hopkins, C. Blanchard, A. Mays, A. Wheeler, JW Nesbitt, J. Robinson, T. Bingham, Pigott, WR Warburton, Ploughman, AE Clarke, WH Harding, J. T. Godfrey, and AEC Blomefield assisted Mr. Phillips.

In 1904 a new pulpit with two tablets affixed was dedicated to Rev Phillips and his wife.

The Coffee Tavern and Institute was built in 1884 on the High-street and cost £1300. It was commissioned as an alternative to the town centre public houses. It is the work of Nottingham based architect Watson Fothergill.

Frank Hemming Kington Fisher was born in 1856 in Salford Parva in Warwickshire. Soon afterwards he is living in Newport and Raglan, Monmouthshire and in the 1871 Census he is still there as a Pupil Teacher. The 1901 Census shows him in Lincolnshire and he eventually becomes a Head Schoolmaster there. His parents and all his surviving siblings were all teachers.

He married in 1878 in Farnham (there are many localities of the same name!). In 1889 he was made a life member of NBKA and qualified as a Bee Expert. He was appointed secretary of the NBKA in 1887 and resigned this position in 1889. During this time he lived in Farnsfield, Nottinghamshire. He died in Sleaford in 1913.

BEEKEEPING IN VICTORIAN NOTTINGHAMSHIRE

Major George Coke Robertson DL, JP (1839 – 1924) was a member of the famous jam making family. He was described as a kindly hearted, generous and religious man. When the Nottinghamshre County Council was formed he was one of the earliest members. In 1870 he changed his surname to Robinson due to his conceived idea that there was a dislike of things Scottish in the area. However, he was Sheriff of Nottingham in 1881 with his original name.

His former residence, Widmerpool Hall (south aspect), is shown above as it was in 2018.

BBJ 15th April 1885

Bligh Competition - Rev TB Garland, Ranby, Retford submitted the following:

June 12th, 1884 placed swarm of 4lbs 11oz in a hive made out of an old gasoline box on 9 frames of wired foundation

June 13th	4 sheets of the foundation drawn but remainder well worked out; added 2 frames
Jun 14th	added 2 frames
June 18th	put on crate with 14 1-lb sections
June 19th	took off 9 sections
August 7th	removed crate of sections, 3 of which were filled; divided the stock and added new queen to the swarm
August 8th	queen, when liberated, flew off, but fortunately settled again amongst her new subjects who received her well
August 9th – 31st	fed and added some frames of foundation
October 14th	prepared for winter but in No 2 found queen cells, some hatched out. Removed all and, hearing queen piping, removed her and added fertile queen
October 16th	liberated the queen
October 18th	found her safe
December 15th	finished packing the bees

March 1st 1885   have not looked at the hive since December

                        BALANCE SHEET
Capital	£2. 0. 0.	Hive, etc.	£0. 8. 9.
9 sections	13. 6.	Bees	18. 9.
		Hive, etc.	9. 9.
		Two queens	6. 0.
		Sundries	8. 1.
		Balance in hand	2. 7.
TOTAL	£2.13. 6.		2.13. 6.

BBJ 15 May 1885

The NBKA now numbers close upon a hundred members in various parts of the county, which we have subdivided into districts and appointed a district secretary to superintend each division. Many of the members have availed themselves of the advice and experience of the district secretaries, who act as experts each for his district. There have already been lectures (illustrated practically) delivered at Farnsfield, Willoughby-on-the-Wolds, Hucknall Torkard, and Mansfield, and each lecture has been attended with good results. Arrangements have been completed to hold the Annual Show in connexion with that of the Notts. Agricultural Society at Lenton, Nottingham, on July 2nd, 3rd, and 4th. A liberal schedule of £20 has been provided, and we hope it will be the means of thoroughly establishing our Association on a proper basis in the county.

Shows in connexion with local flower shows have been also arranged at Hucknall Torkard, Farnsfield, Mansfield, Willoughby, Retford, and Radcliffe-on-Trent. The bee-tent will be in attendance at all. Members will see from this that we have not been idle, and that we had plenty to do to revive the Association which all but died last year. I enclose a newspaper cutting of a report in the Notts. Express of April 24th, furnished by one of our district secretaries of his visit to a part of the county, where bee-keeping is little known on the bar-frame principle.

E Ferneyhough, Honorary secretary.

A Visit to Keyworth Beekeepers. One of the many privileges of the members of the above Association is the visit of an expert at least once a-year to their apiaries, for the purpose of giving them instructions and assistance in the management of their bees, the charge in no case to exceed the cost of railway fare. The Rev. HP Ling, rector of Keyworth, being a member, availed himself of the opportunity, and as it is my duty, being one of the district secretaries, I had much pleasure in going and explaining the modern methods of managing bees for the largest profit. I was agreeably surprised when I arrived on Tuesday afternoon to find Mr. Ling had assembled in his garden a large number of his villagers, who keep bees, to witness the manipulating of his bar-frame (as there is none in use in the village except his), and to hear and see the advantage of the modern system of bee-keeping, and the benefits of becoming Members of the NBKA.

After I had explained the advantages of the bar-frame, and how much more control they had over their bees in them, and how easily they could ascertain all that the bees were doing inside the hive, by taking out the frames when the weather admitted them so doing, and a better way of obtaining the honey than that of stifling the poor insects with fumes of brimstone, and cutting out of the hive a confused mass of comb, pollen, grubs, and honey contaminated with sulphur and covered with the dead bodies of the victims. We then adjourned to the gardens of some of the cottagers who keep bees, to examine theirs in skeps, and to show them the way of getting super honey from them, as it is worth considerably more money per pound than drained honey.

Keyworth is a very good part of the county for keeping bees, as there is an abundance of honey-yielding trees and plants: all the hives have plenty of stores and are strong with bees, except one. The man that pointed it out to me said he could always tell which was the strongest, although he never lifted them off the stand (except to put them on the sulphur-pit), and, strange to say, the hive he pointed out to be the strongest had no bees in it. What made him think it was a strong one was that the bees from the other hives were clearing all the honey out of it. The cause of the stock dying out was no doubt that the queen died last autumn, it being too late to rear another in her place, as there was plenty of honey left in the hive.

On Monday evening, May 4th, at St. John's Schoolroom, Mansfield, a lecture was given by Mr. FKH Fisher, of Farnsfield, showing the advanced system of keeping bees and gathering honey. The Rev. W Maples occupied the chair. The lecture was given at the monthly meeting of the Mansfield Horticultural Society, at who's Annual Flower Show an exhibition of bees, honey, hives, &c, will be held. The lecturer showed in a clear and able manner how bees may be kept to be a source of profit and interest. Practical illustrations were appropriately introduced, the audience being deeply interested in the subject, which was made intelligible to the veriest novice. A vote of thanks to the lecturer and the chairman closed the proceedings.

Mansfield Horticultural Society was founded in 1876. In 1894 it was described as doing good work amongst the working classes of the town and district by means of its annual exhibition held on August Bank Holiday.

BBJ 15th May 1885
WHH We have been informed that the iron wire mentioned is sold by Mr Walton, Muskham, Newark.

Sheffield Daily Telegraph 8 Jun 1885
The Mayor of Retford, Mr Henderson, presided at a meeting on Saturday to advance the culture of bees in the neighbourhood. Mr FHK Fisher, of Farnsfield, delivered an address, and it was shown that the work might be extended and carried on with great profit.

NEP 2 Jul 1885
The eighth formal exhibition of cattle, horses, sheep, poultry, &c, in connection with the Nottinghamshire Agricultural Society was opened to the public this morning, and will remain open during to-morrow and Saturday.
Bees
Class 1 — for the best specimens of Ligurian, Carrniolan, Cyprian, Syrian Honey Bees, be exhibited with the queen in Observatory Hive — 1, EC Walton, Muskham
Class 2 — for the best specimens English Bees, to be exhibited with the Queen in an Observatory Hive — 2, EC Walton
Honey
Class 3 — for the largest and best exhibition of super honey the produce of one apiary during the year 1886 — 2, Geo. Coope, Farnsfield
Class 4 — for the largest and best exhibition of extracted run honey in glass jars, each jar to contain one or two pounds net weight of honey respectively, the produce of one apiary during the year 1886 — 2, Geo. Coope
Classs 5 — for the best twelve 1 lb. sections of comb honey, in crate — 1, Geo. Coope. 2. John Beeson, Lamcote, Radclíffe-on-Trent
Class 6 — for the beet twelve 1 lb. glass jars of extracted or run honey — 1, Mrs. Wotton, Widmerpool; 3, Frank HK Fisher, Farnsfield
Class 7 — for the best super comb honey — 2, G. Caparn, Newton, Nottingham
Hives
Class 8 — for the best complete and most practical hive the movable comb principle, five of the frames to be fitted with comb foundation, with arrangements both for storing surplus honey and for wintering. Price not to exceed 80s — 2, EC Walton, Muskham
Class 10 — For the cheapest, neatest, and best super for harvesting honey in the comb in a saleable form — 2, EC Walton
Class 11 — For the best and most complete collection of hives and bee furniture most applicable to modern bee-keeping. No two articles to be alike — 2, EC Walton

George E Caparn was listed as a farmer's son in the 1881 Census and was born in 1859 in North Witham, Lincolnshire.

NEP 3 Jul 1885
The eighth exhibition of the Notts. Agricultural Society, which was opened yesterday in the spacious enclosure at Lenton, evidences a small but significant increase in the

number of entries, and the show can be said to bear very favourable comparison, with those of previous years. The exhibition of bees, under the auspices of the NBKA, was well advised addition to the exhibition, and has not failed to secure considerable attention.

NG 24 Jul 1885
Hucknall-Torkard Flower Show: The tent of the NBKA was erected on the ground. Several hives of bees were driven, the queen bee captured, and other methods of manipulation were practically exhibited. In the bee show the following awards were made:
Stock of bees, of any race to be living, with their queen in an observatory hive – A Simpson
Six one pound sections of honey –A Simpson
Three two pound sections of honey – W Bell
Run Honey, in glass jars, not less than six pounds, to be shown – 1. Simpson: 2. Littlebury
Bell glass of comb honey – S Vickers
Special section honey – W Bell

NEP 31 Jul 1885
The annual flower show of the Nottinghamshire Horticultural and Botanical Society was commenced on Thursday in Mapperley Park, Carrington, which was kindly lent for the occasion by Aid. Lambert, JP. Contrary to the general experience of the society the weather was beautifully fine, and as a result there was a large and fashionable attendance.
Bees, Honey and Hives
Bees— Class A— Specimens of bees, to be exhibited with the queen in an observatory hive— Mr. EC Walton, Muskham
Honey - Class B— Twelve 1-lb. sections of comb borne nome (?) in crate— Mr. Cooper, Farnsfield
Class C, -Twelve 1-lb. glass Jars of extracted or run honey - Mr Wootton, Widmerpool
Hives— Class D - For complete and practical hive on the movable comb principle, five of the frames to be fitted with comb foundation, with arrangements both for storing surplus honey and for wintering— Messrs. Turner and Son, Radciiffe-on-Trent
Class E— For complete and practical hive on the movable hive principle, with arrangements for storing surplus honey on smaller scale— Mr. EC Walton
Class F— For complete collection of hives and bee furniture most applicable to modern bee-keeping, no two articles to be alike— Messrs. Turner and Son

The bee and honey exhibition was a very good one, but the appliances for hives were only a moderate display. Practical illustrations were given of manipulating with live bees, showing the best methods of driving, capturing the queen, transferring comb from straw skeps to bar frame hive, uniting stocks. Arrangements were made by which visitors might view with safety the mysteries of the hive and witness the perfect command the scientific apiarist has over his bees. Demonstrations in advanced straw hive management were also made with a view to show how needless is the cruel practice of killing bees to obtain honey, even under the straw hive system of beekeeping.

NG 28 Aug 1885
Retford Flower Show. The annual exhibition in connection with the Retford and District Gardening Society took place on Wednesday. The weather was favourable, and owing, probably, to extra attractions having been introduced, the attendance was somewhat larger than usual. An attractive feature, was an exhibition of bees and bee-keeping appliances. Mr. W. Silver, local secretary of the NBKA, several times during the day gave exhibitions of bee driving, manipulation of hives, &c

NEP 18 Sep 1885
Honey Fair. The committee of the NBKA hold their first annual honey fair at Mr. Laslett's Auction Room, Market-street, to-day and to-morrow, the idea being to find a ready market for the sale of their members' harvest. About two tons of honey is staged, and should the sale be a success it will prove of great benefit to the cottagers and others who have sent their summer's harvest.

In 1880 this company traded as Warwick and Laslett. The following notice was published in the local press in 1885, "TJ Laslett, auctioneer and valuer (late of the firm of Warwick and Laslett has commenced business on his own account and has secured the local and largest Auction Rooms. By 1885 he was back to 21, Market-street. He died in April 1888 aged 46.

NG 18 Sep 1885
One of the most interesting features in connection with the exhibition was the annual show of the Derbyshire BKA. The classes for honey were well filled, the cottagers competing in good numbers. The BBKA silver medal was taken by Mr J Gower, of Hucknall, near Mansfield, with an exhibit of superior quality weighing between 60 and 70lb. The winners of the first prizes in the cottagers' classes included Mr. A. Simpson, of Mansfield Woodhouse.

NEP 19 Sep 1885
The NBKA held their first annual honey fair yesterday, at Mr. TJ Laslett's auction rooms in Market-street, Nottingham; and judging from the large quantity of honey (about two tons) exhibited and the prices obtained during the day - the average price being 1s 2d per lb - the show promises every respect to be a great success.

The BBKA gave a silver medal and also a bronze medal for competition, the former for the best six sections of comb, which was awarded to the Rev. JB Garland, of Ranby, Retford; and the latter for the best six jars of extracted honey, which went to Mr. A. Felstead, of Rempstone. The entries in both classes were numerous, and the competition very keen. The most striking feature of the show, however, is a bell glass super, weighing about 40 lb, sent by Mr. GE Caparn, of Newton. Great credit is due to the secretary, Mr. Ferneyhough, of Radford, and the committee, for what they have done towards bringing the affair to such a successful issue. The fair will be continued to-day.

BBJ 14 Jan 1886
The Annual General Meeting of NBKA will be held at the People's Hall, Heathcote-street, Nottingham, on Saturday, Feb. 20th, at 3pm

BBJ 18 Feb 1886
Question – Extracting. By all means extract indoors, make sure the hives you extract from are securely covered down before leaving them, or it will cause very much fighting. Why not build a cheap garden-house?

Thomas Rose, Lamcote, Radcliffe-on- Trent

NEP 20 Feb 1886
The first annual meeting of the members of NBKA was held this afternoon the People's Hall, Heathcote-street. The Rev. JD Phillips presided, and there was moderate attendance. A letter was read from Lord Newark, the president of the association, expressing his regret at not being able to attend, and enclosing a subscription of £1. The Hon. Sec. (Mr. Ferneyhough), presenting the annual report, expressed his regret that the association had not succeeded quite so well it had been hoped would have been the case. Taking all things into consideration, however, especially the fact that was the year of their apprenticeship, they had not done badly, though there was balance of £9 13s 5d. on the wrong side of the account. The subscriptions for 134 members and the donations amounted to £24 14s 8d. The bee tent that the shows had involved showed a loss of 13s 11d. The expenses of the lectures had been £4 8s, and the honey fair had resulted loss of £10 7s. The association was perfectly solvent, however, because with the subscriptions for the present year there was still a small balance in hand. Mr. Felstead proposed, and Mr. Godfrey seconded, the adoption of the report, which was unanimously agreed to.

Lord Newark was re-appointed president, and Ald. Manning, Mrs, Robertson (Widmerpool), and Mrs. Hole (Caunton) were re-elected vice-presidents.

The committee, was agreed, should consist of the same ladies and gentlemen as the previous year, with a few exceptions, and Mr. Ferneyhough was appointed honorary secretary and treasurer.

It was agreed that the bee tent should be hired from the BBKA, for use at the shows during the coming year, the chairman observing that he thought the tent tended to advertise the association in a way which could not otherwise the case, this opinion being shared the majority of those present. The drawing for the best hives next took place, and resulted in Mr. RS Aslin, Carlton, securing the first prize, and Mr. Richard Turner, of Radcliffe-on-Trent, the second; the latter being a maker relinquished the prize, which was awarded to Mr. HE Hollins, of Mansfield.

After the transaction of some other business of an unimportant public character, the meeting closed with a vote thanks to the chairman.

BBJ 23 Feb 1886
How to Commence Beekeeping - Samuel Kirkby, Beeston. 16 Feb

If a start in beekeeping you're anxious to make, the following hints may
be useful to take
That from the pursuit may arise satisfaction instead of disgust with your stocks,
and distraction;
For bees when mismanaged are sources of danger, and spare
with their sting neither owner nor stranger.
Get a good strong frame-hive in the spring, for this reason,
the paint can get dry, and the wood can well season;
And in view of your harvest you'll have less vexation if you have
standard frames filled with good comb foundation.
The first hive you buy should be from a good maker, Abbott,
Neighbour, Blow, Baldwin, Edey, Howard, or Baker,
Or any good firm whom you see advertise will be glad to supply a
good hive and advise.
Then with veil, gloves, and smoker, you need have no fear of the
first four pound swarm you have fortune to hear of;
To get this in May will repay extra trouble, compared with
July it is fully worth double.
And if possible try by all means to contrive it to have it brought home
the same day that they hive it,
Say an hour before dusk; place your hive on the ground,
prop it open an inch, then a sheet must be found
Draw this well up the floor-board, place quilts on and feeder,
throw the bees on the sheet and they follow their leader.
To see them march in is most truly delightful: If vicious,
your smoker will make them less spiteful.
They soon hasten in; close the hive about dark,
lift the bees to their stand and they're ready for work.
They draw the foundation, and very slight feeding will stimulate better

> for working and breeding.
> And now if the weather be clear, warm, and sunny,
> your hive will soon fill with the purest of honey.
> If further advice of the best you would seek,
> consult the Bee Journal that comes once a-week:
> And there you will certainly find the true key to pleasure and profit
> from our honey bee.

Samuel Kirkby was born in Beeston in 1849. In 1874 he married Annie Thornhill. He was a talented pianist and choirmaster. He shared the duties of organist with his brother Frederick. At the time he wrote the above article he would have been 37 years old. He died in 1916 in Beeston.

BBJ 29 Apr 1886
Received from Mr. EC Walton, of Muskham, his Illustrated and Descriptive Price List of Moveable Comb Bee-hives and Apicultural Appliances (48 pp.)

BBJ 6 May 1886
Exhibitors at Shows. Are we poor would-be exhibitors to understand that to have a chance of winning a prize at the BBKA's show we must glaze each separate section on both sides, and then put it into a show-crate having glass on each side as well, or (what a boon!) that we can get out of the difficulty by first putting each section into a small box having glass sides, and then place box in a show-crate? Why this double doing it? What does uniformity gain by it? Would not a plain show-crate showing through glass sides each side of every section please the eye and keep robber bees at bay as well? Or, if I may be so bold, that each separate section must be glazed on both sides if not shown in a (glass-sided) crate that shows each side of every section without unpacking?

The quarter-inch paper edging is not much certainly, but just enough to hide many defects or pop-holes in what may otherwise be a perfect section, and would lead to sections taking prizes before those which perhaps were not quite so good in the centre of the visible surface, but which might have no other defect, and would therefore be better sections than those taking prizes before them.

Then the expense. I have always understood that one of the rules of our glorious BBKA is that beekeeping was to be taught more especially for the good of the poorer or cottager class. Yet we have here a rule which will put our cottagers outside the exhibition altogether, and, in a sense, others besides cottagers too. I for one cannot afford to keep bees unless at a profit. I got only an average of nine pence for my sections last season, and shall never get more, and most likely less; then why should we go in for unnecessaries?

As to bottled honey does the rule mean that the bottles must be corked and capped in addition, or will the cork inside the screw-caps meet the rule, and that the bottles should only be corked when parchment, etc, are used for covering instead of screw-caps? Friar Tuck, Mansfield, Woodhouse, Notts.

Hants and IoW Show June 24th 1886
The following is the list of awards to EC Walton, Newark:
For collections 3rd; observatory hives 1st: 35s hives — 3rd: 10s 6d. hives—2nd: extractors—3rd: skep and racks—2nd

NEP 28 Jul 1886
Farnsfield Flower Show. NBKA offered prizes for honey to persons residing in the parishes of Farnsfield, Halam, Edingley, Blidworth, Oxton, Kirklington, Eakring, Maplebeck, Rufford, Bilsthorpe, Southwell, Farndon, and Upton

Leicester Chronicle 31 Jul 1886
Leicestershire Agricultural Society.
Frame hive – 1st, EC Walton, Muskham, Newark
Straw skep – 2nd, EC Walton

Sheffield Daily Telegraph 4 Aug 1886
Great Yorkshire Show. Not the least interesting portion of the exhibition was the tent in which the bee-keeping appliances are grouped. A new reversible crate is shown among a large number of other appliances by Mr. EC Walton.
Prize list:
Bar-frame hive, not exceeding 16s First prize. £1, EC Walton. North Muskham. Newark
Bar-frame hive, not exceeding 10s First prize, £1, EC Walton
Straw Hive, flat-topped, with hole not less than three inches wide, Second prize, 10s, EC Walton
Best exhibit Bee furniture Second prize, 10s, EC Walton
Six samples of run or extracted honey, in 1 lb jars of clear plain glass. Second prize. 5s EC Walton

BBJ 12 Aug 1886
The annual show of Leicestershire BKA took place at Leicester on July 28th and 29th 1886. There were ninety entries, but of these only sixty-three were hives, &c.
VI. For the best frame hive, price not to exceed 10s 6d; 15s, 10s, 5s. Five entries. First, EC Walton, Muskham
VIII. For the best straw skep, with flat top and arrangements for sectional supering, 7s 6d, 5s. Five entries second, EC Walton, Muskham

BEEKEEPING IN VICTORIAN NOTTINGHAMSHIRE

BBJ 9 Sep 1886
Prices of Hives Exhibited at Shows. I notice with some pleasure, and most fully endorse, the remarks made by Mr. L. Wren in a recent Journal on a subject that wants well ventilating in the Bee Journal. I am glad to find someone has at last had courage to open the ball on the matter.

My experience (which is not very limited) of shows quite coincides with Mr Wren's as regards hives that are exhibited for competition at scheduled prices, it being perfectly evident that some so exhibited cannot be produced for the money. For instance, I happened to attend this season a show in the south of England, and spoke to the exhibitor of the first prize hive in company with another exhibitor, asking him how he made the hive for the money. The answer I received at once showed me the utter futility of attempting to gain honours by showing an article which could be produced at the price specified. Surely this is a matter that ought to be most carefully investigated, as it is manifest to all that it causes much dissatisfaction between manufacturer and customer, the latter expecting an article that cannot be produced for the sum specified. EC Walton, Muskham, Newark.

BBJ 30 Sep 1886
Derbyshire BKA. An exhibition of bees, honey, and appliances was held in a tent on the Recreation Ground on September 8th and 9th, 1880. The following prizes were awarded to Mr A Simpson, Mansfield Woodhouse

> Bees, of any race, in Observatory hives: 1st
> Super honey (sectional or otherwise), the produce of one apiary, 1886: 2nd
> For the best twelve sections: 2nd
> Run honey, in class jars 2nd
> Bees' Wax. 2nd
> Bee Driving. Class 1, open to all comers, 1st

Amongst the judges was Mr. EC Walton, of North Muskham, Newark.

BBJ 25 Nov 1886
Novelties.
We have received from Mr. EC Walton, North Muskham, a specimen of a wicker covered jar similar to the ones exhibited by him at the Norwich Show, and in which the honey was presented to HRH the Princess of Wales. The one before us is a great improvement on those, in that the jars are made to slip into the wickerwork covering and can be easily removed for the purpose of cleaning them. It is also provided with a light handle, so that a purchaser can carry away the jar of honey without being obliged to wrap it up in paper. We are also pleased to find that the prices have been very much reduced.

They can be made to fit any sized bottle, and are both useful and ornamental. On page 542 of our last volume we described a wicker-work covering for honey jars which was introduced by Mr. EC Walton, Muskham, Newark, at the Norwich Show; and we are now able to give an illustration of it which the inventor has been good enough to send us. The cover is made to suit any sized jar, and it will be seen that by slipping up the ring the jar can be easily removed for cleaning purposes.

BBJ 30 Dec 1886
Business Directory. BBJ 1887
For the use of Manufacturers and Purchasers of Beekeeping Appliances. Hives and Other Appliances. Honey Merchants. Foreign Bees and Queens. Metal Ends. WB Baker, Muskham, Newark

---

The New Patent Honey & Fruit Squeezer.

SPECIAL NOTICE! — In consequence of the increased demand for this useful article, the Manufacturer has enlarged his plant, and is now able to produce them at a much lower cost. Price 2s. 6d. each, post free 2s. 9d.

For particulars see 'British Bee Journal,' page 382.

Agents for the United Kingdom,

TURNER & SON, BEE-HIVE MANUFACTURERS, RADCLIFFE-ON-TRENT, NOTTS.

NEP 30 Apr 1887
The annual meeting of the NBKA was held at the People's Hall, Heathcote-street this afternoon. Mr. P Pilgrim, Shelford, presided, and there were also present Mrs. Wootton, Mr. CP Brearley, Mr. Felstead, Mr. Simpson, Mr. EC Walton, Mr J Colgrave, Mr WPJ Allsebrook, &c. Mr FHK Fisher, Radcliffe-on-Trent (hon. secretary *pro. tem.*), said Mr. E. Ferneyhough, the hon. secretary, was unable through press of business to fulfil his secretarial duties, and had asked him temporarily take his place. The balance sheet was not ready, but they had considerably reduced the adverse balance of last year. During the year the Association had given a few prizes at the Mansfield Agricultural Show and also other shows. With regard to the proposed honey fair at the latter end of the season, nothing had been done, neither had any association business been transacted, nor had any committee meetings been called. Many people were anxious that the association should carry on and he thought they should make a strong feature of the honey fair.

The chief business of to-day was the election of officers and new hon. secretary.

On the proposition of Mr Walton, seconded by Mr. Colgrave, and supported by Mr. Allsebrook, Mr. Fisher was appointed hon. secretary and treasurer, an honorarium of £5 per year be presented to him by the association. Lord Newark was chosen president, and the other officers were elected as follows: Vice-presidents: Lord Charles Bentinck, Mrs. Robertson, Widmerpool Hall, and Mrs. Hole, Caunton Manor. Committee: The Rev. TB. Garland, Rev. HP Ling, Mr. A. Felstead, Mr. Thos. Rose, Mr Wm. Silver, Mrs. Wootton, Mr. Marriott, Mr. Brearley (Carlton), Mr. Godfrey (Langley), Mr. Gosling (Arnold), Mr. R. Turner (Radcliffe), Mr. E. Ferneyhough, and the district secretaries. Manipulator and lecturer at the Shows, in connection with the Society, Mr. EC Walton, North Muskham. A vote of thanks was accorded to Mr. E. Ferneyhough for the services he had rendered the Association, and the Chairman having been thanked, the meeting terminated.

Lord Charles Cavendish-Bentinck DL, JP, DSO born 1868 and died 1956 at Oxton Hall. He fought in both the Boer War and WW1 when he was mentioned in despatches and wounded several times in both.

Thomas J Gosling was born in Bulwell in 1818. He was shown as Relieving Officer who was an official appointed by a parish or union to administer relief to the poor.

NEP 4 May 1887
We are requested state that the address of Mr. Frank HK Fisher, hon. sec. of the NBKA, is Farnsfield, near Southwell, and not Radcliffe-on-Trent.

BBJ 12 May 1887
The annual meeting of NBKA was held at the People's Hall, Heathcote-street, Nottingham, on Saturday, May 1. Mr. P Pilgrim, of Shelford, presided, and amongst those present were Messrs. Brearley, A Felstead, A Simpson (Mansfield Woodhouse), EC Walton (North Muskham), J Colgrave, Allsebrook, Turner, and Marriott. Mr. FHK Fisher, hon. secretary *pro tem.*, said that he had been asked to officiate for Mr. E. Ferneyhough, who through pressure of business had been unable to attend to the work incident to his acceptation of the post of honorary secretary and treasurer. He (Mr. Ferneyhough) was not able to be present at the meeting, and had not as yet prepared the statement of accounts. There was still a balance on the wrong side, but it was considerably less than that of last year. Up to some short time ago he was in hopes that there would be a balance in hand, because so little had been done by the Association last year. They had a great many members —109—last year, and the only business was that done at the Mansfield Agricultural Show, where they had a bee tent, and a few prizes in connexion with other shows.

Messrs. Morris and Place last year offered the Association the use of their mart for the purpose of holding a honey fair but this matter had fallen through. The prizes won at the Mansfield Agricultural Show were in the hands of Mr. Barron. The holding of a honey fair ought to be one of the points to be taken up strongly by the committee this year if they went on with it. Members stated that the chief difficulty they had in connexion with beekeeping was getting rid of their honey. If a fair was held this would be done away with. In conclusion, Mr. Fisher stated that Mr. Ferneyhough (Radcliffe-on-Trent) had promised to get his accounts properly made out in a few days.

Mr. FHK Fisher (Farnsfield) was appointed secretary, Lord Newark president, and Lord Charles Bentinck, Mrs. Robertson (Widmerpool Hall), and Mrs. Hole (Caunton Manor), vice-presidents. Mr. EC Walton, of North Muskham, was re-appointed manipulator and lecturer at shows held in connexion with the society.

The following were appointed on the committee for the ensuing year: Revs. TB Garland, and HP Ling; Messrs. A Felstead, T Rose, W Silver, Mrs. Wootton, Marriott (Nottingham), Brearley (Carlton), Godfrey (Langley), Gosling (Arnold), R Turner (Radcliffe-on-Trent), and E Ferneyhough (Radcliffe).

A vote of thanks was accorded Mr. E. Ferneyhough for his services as honorary secretary and treasurer, and a similar compliment to the chairman closed the meeting.

NEP 22 Jul 1887
Mansfield Flower Show. The first prize in the open class for extracted honey this show was carried off by Mr. Frank HK Fisher, the honorary secretary of the NBKA, and not by Mr. D. Bowler, as stated in our report.

Sheffield Daily Telegraph 16 Sep 1887
The annual exhibition of bees, honey, and appliances, under the management of the Derbyshire BKA, was the largest and best display of this character ever held in this district, and fully proved the great progress which bee culture has made of recent years. The specimens of honey by Mr A Simpson, of Mansfield Woodhouse and others, were of the finest flavour and description. The judges were Mr. TB Blow, Welwyn, Herts, and Mr. EC Walton, Newark.

BBJ 4 Oct 1887
Prize-takers at Shows. It is generally understood that the gainers of prizes at shows undertake to supply the general public with the articles exhibited, such as hives, feeders, &c, at the prices stated by them at the show. We are not aware that the Royal Show at Nottingham is an exception to this acknowledged rule.

Prizes for honey open to members of BBKA are offered for competition at the Annual Show of NBKA to be held at Sutton-in-Ashfield, on July 25th, entries for this Exhibition close on July 9th. Application for schedules to be made to Mr FHK Fisher, Farnsfield, Southwell, Notts.

BBJ 4 Oct 1887
The Affairs of a North Muskham Manufacturer.
A first meeting of the creditors in the case of William Burton Baker, of North Muskham, manufacturer of apiarian appliances, under a receiving order dated August 22nd, was recently held at the offices of the Official Receiver, No. 1 High Pavement, Nottingham. In the absence of Mr. Thorpe the Deputy Official Receiver (Mr JYV Jeffries) presided. Mr. Grosvenor Hodgkinson, of Newark, appeared for the debtor and produced a medical certificate showing that Mr. Baker was unable to attend in consequence of illness.

Creditors were represented by Mr. Metcalfe, of Southwell, and Mr. Robert White, of Newark. Proofs having been admitted, the Deputy Official Receiver said that the debtor's statement of affairs was not lodged until that morning and had not yet been sworn to. Mr. Hodgkinson had produced a certificate as to the debtor's illness, which he presumed was the cause of the delay. Mr. Hodgkinson said that was so. He had prepared the statement from papers supplied to him by the debtor. The Deputy Official Receiver said that the gross liabilities, according to the statement, were £5236 8s 2d; the amount to rank for dividend being £5047 8s 2d. That did not include a claim for a large sum made by the trustees under the marriage settlement—a claim which would require investigation before being admitted for dividend. The assets as estimated by the debtor were £1585 7s 10d from which would have to be deducted 64/- for preferential claims, leaving a balance of £1521 7s 10d, or a deficiency of £3526 0s 4d. The debtor accounted for the deficiency in this way. He said that his excess of liabilities on August 27th, 1887, was £2451 0s 4d; his net loss in carrying on business for the past year had been £500; bad debts amounted to £50, and household expenses for the year £300. In addition to that he had paid as surety for a brother £225, making a total altogether of £3526 0s 4d. There had been nothing realized from the estate at present.

The debtor having no offer to make to his creditors had been adjudicated bankrupt, and the case being a non-summary one the estate would have to be administered by a trustee. Replying to a question, Mr. Hodgkinson said that he hoped to get the statement of affairs sworn to shortly. Mr. Robert White was chosen as trustee, with a committee of inspection consisting of three creditors.

BBJ 3 Nov 1887
The NBKA committee are working on steadily to promote the interests of beekeepers in the county. The bee-tent has been sent and prizes given at three shows, viz. Retford, Sutton-in-Ashfield and Farnsfield. Bee publications are circulated monthly

to all the members and the district secretaries have been busy giving advice and assistance to them.

Messrs. Morris & Place, auctioneers, Bridlesmith Gate, Nottingham, having kindly offered the use of their auction mart to the committee it is intended to hold a honey fair there on Friday and Saturday, December 16 and 17. At this fair the silver and bronze medals and certificate of the BBKA (with which the NBKA is affiliated) will be offered for competition.

Full particulars will be advertised in due course. One of the aims of the Association is to introduce the best form of hive to bee-keepers and to attain this hives are drawn for at the annual meeting. For the drawing next January several manufacturers and others have promised hives and appliances.

Some districts of the county are without secretaries and the committee are anxious to get ladies or gentlemen to take the posts before the end of the year, so that their names may be printed in the annual report. Lectures will be given on bee-keeping throughout the county during the winter whenever desired, application be made to the Honorary Secretary, Mr. FHK Fisher, Farnsfield, Southwell.

BBJ 8 Dec 1887
While thanking our Editor for his courtesy in admitting our junior readers to compete for prizes by solving our problem we must ask permission to trespass a little further on his space in order to give a list of competitors who were successful with their poetical solutions. The prizes—for we now propose to give three—have been adjudicated as follows:
>Equal first prize of a frame hive SP Kirkby, Beeston.

Although Mr. Lewis is beyond the age specified we have great pleasure in offering him a hive, if he will do us the honour to accept it for his most excellent solution. Will he oblige us by forwarding his address?

>As you have offered boys a hive, to get it, sir, I think I'll strive,
>And father's promised, if I win it, to put for me a swarm within it.
>Six tomtits, then, sir, if you please, will in six minutes kill six bees;
>It follows, then, clear as the sun, that in one minute they kill one.
>In fifty minutes thus they will just manage fifty bees to kill;
>And in this time, if I've not blundered, twelve tomtits, sir, would kill one hundred.
>And now I've done my very best, in faith I'll leave to you the rest.
>My age is twelve, and, if I've won it, here's my address to put upon it.
>>Samuel P Kirkby, Imperial Park, Beeston.

BEEKEEPING IN VICTORIAN NOTTINGHAMSHIRE

NEP 16 Dec 1887
Notts. Honey Fair. After a lapse of two years the honey fair in connection with the NBKA was revived to-day with encouraging prospects. Messrs. Morris and Place kindly lent their auction mart in Bridlesmith-gate, Nottingham, for the purpose of the display and very considerable interest was shown the proceedings by those interested in the work of the association. Altogether more than a ton of honey was staged and the quality was highly satisfactory, there being over 20 exhibitors. The BBKA offered a silver medal for the best six 1-lb. sections, a bronze medal for six 1-lb. jars of extracted honey, and a certificate for a bee-glass of honey, for which there was a keen competition. The judging took place last night, the awards being made by Mr. Henry Yates of Grantham, and Mr. Godfrey, Lambley. Mr William Silver, Bridge-gate, (Retford), won the principal medal with some excellent sections. The bronze medal went to the Rev. RA McKee, of Farnsfield, and the certificate Mr. G Caparn, of Newton, who exhibited a bell glass of 75 lb. of excellent honey. The arrangements have been efficiently carried out by Mr. FHK Fisher, Farnsfield, the honorary secretary of the association, of which Lord Newark is president. The sale being conducted to-day by Messrs. Brearley and Turner, under the superintendence of the secretary. Up to this evening few sales had been effected. The prices asked ranged from 1s to 1s 6d per lb. Wax, of which there was a small supply, was selling at 1s 8d per lb. The exact quantity of honey staged was 2,073 lb. and of wax 13 lb. The fair will be continued tomorrow.

Robert Alexander McKee, MA, was born in Ireland in 1847. Little is known of his time in his first parish of Lumb in Ireland, but his successor wrote "Coming immediately after Mr. McKee, I can myself testify to the excellent condition in which I found the affairs of church, schools, vicarage house and parish at large and to the good church tone which prevailed."

He was vicar of St Michael's church, Farnsfield from 1882 until 1922. There is a memorial tablet to his service to this parish on the south wall of the chancel in the church.

The bronze medal won by Rev Mckee shown here is currently (2018) in the possession of Maurice Jordan, one-time secretary of NBKA.

Sheffield Daily Telegraph 19 Dec 1887
NBKA. The annual Honey Fair was held on Friday and Saturday in Nottingham.

BBJ 5 Jan 1888
After a lapse of two years the honey fair in connexion with NBKA was revived on Friday and Saturday, December 16 and 17, with encouraging prospects. Messrs. Morris and

Place kindly lent their auction mart in Bridlesmith Gate, Nottingham, for the purpose of the display and very considerable interest was shown in the proceedings by those interested in the work of the Association. There were over twenty exhibitors.

The BBKA offered a silver medal for the best six 1-lb sections, a bronze medal for six 1-lb jars of extracted honey and a certificate for a bee glass of honey, for which there was a keen competition. The judging took place on Thursday night, the awards being made by Mr. Henry Yates, of Grantham, and Mr. Godfrey, of Lambley. Mr William Silver, of Bridge Gate, Retford, won the principal medal with some excellent sections. The bronze medal went to the Rev. RA McKee, of Farnsfield, and the certificate to Mr. G. Caparn, of Newton, who exhibited a bell-glass of 75 lbs. of excellent honey.

The arrangements have been efficiently carried out by Mr. FHK Fisher, of Farnsfield, the honorary secretary of the Association. Wax, of which there was a small supply, was selling at 1s 8d per lb. The exact quantity of honey staged was 2073 lbs and of wax $13^{1}/_{2}$ lbs. A comparatively small quantity of the honey staged was sold, considerably more than half remaining. Prices, which on the first day were quoted at 1s and 1s 6d per lb., fell to as low a rate as 8d, 9d, and 10d, both for run and comb honey retail, and 6d wholesale.

BBJ 5 Jan 1888
The BBJ is published by Kent & Co., 23 Paternoster Row, and may be obtained of all local Booksellers, and of the following Agent: Baker, WB, Muskham, Newark.
HONEY MERCHANTS. Baker, WB, Muskham, Newark.
FOREIGN BEES AND QUEENS. Baker, WB, Muskham, Newark.
METAL ENDS. Baker, WB, Muskham, Newark.

BBJ 26 Jan 1888
For Sale. New Vertical Engine, suitable for Driving' Circular Saw. Cheap. Crankshaft and Fly-wheels. Address A. Green, Selston, Notts.
For Sale. Treadle Saw Bench, great power and speed. Cheap. Particulars of A. Green, Selston, Notts.

NEP 30 Jan 1888
The annual general meeting of NBKA was held on Saturday at the People's Hall, Nottingham. Mr J Gosling was elected to preside, and amongst those present were Mrs. Wootton, Messrs. P. Scattergood, jun. R Turner, S Godfrey, D. Burnham, HJ Raven, SW Marriott, and FHK Fisher. Mr. Fisher (hon. secretary) read letters apologising for absence from Lord Newark, MP, Lord Charles Bentinck, and others. Mr. Fisher also presented the report for 1887, which stated that they had every reason to be satisfied with the results of the year. The Agricultural Society had an account against the Beekeepers' Association for the shows of 1885 and 1886, which

the state of their funds had not allowed them to settle. There were other unpaid debts contracted before 1887. Last year's expenses had been met, although the membership had fallen off.

With increase of members there was every prospect of the association paying off its debts. The decrease of the number of members was due the fact that in 1886 they got little nothing for their subscriptions. The honey fair was a loss of £8 to the association. The committee would be glad to arrange for lectures in different parts of the county if desired. Mr. EC Walton, of North Muskham, had resigned the office of expert in consequence of pressure of business. The balance-sheet for the year 1887 showed that the income had been £23 11s $2^{1}/_{2}$d., and the expenditure £34 1s 6d. The report and balance-sheet were adopted.

Votes of thanks were accorded to the retiring office-bearers and to Messrs. Morris and Place for the use of their mart on the occasion of the honey fair. The office-bearers for the ensuing year were elected, the president being Lord Newark, MP.

The annual drawing for bee-keeping appliances, which had been presented various manufacturers, proceeded, with the following result—Hives, Mr. S White (Bleasby), Mr Chadwick (Mansfield), Lord Charles and Mr WC Measures (Upton); comb-box, Mr. F Fisher; feeder, Miss H Parkyns (Woodborough); smokers, Mr. Joseph (Upton) and Mr. G Coope (Farnsfield); extractor, Mr. F Pickard (Mansfield); books, Mr WF Newman (Calverton) and Mr. G Fisher (Farndon), 200 1-lb. sections, Mr J Wilson and apifuge, Mr. TJ Gosling (Arnold).

Miss Hilda Parkyns born in 1863 was the last of eight daughters of Mansfield Parkyns of Woodborough Hall.

BBJ 2 Feb 1888
The annual general meeting of the NBKA was held on Saturday at the People's Hall, Heathcote-street, Nottingham. Mr. TJ Gosling was elected to preside, and amongst those present were Mrs. Wootton, Messrs. P Scattergood, Jun., R Turner, S Godfrey, D Burnham, HJ Raven, S Marriott and FHK Fisher.

Mr. Fisher (hon. secretary) presented the annual report for the year 1887, which stated that they had every reason to be satisfied with the results of the year. The committee had met six times and had done all they could towards promoting bee-keeping and assisting beekeepers. Last year's expenses had been met, although the membership had fallen off. With an increase of members there was every prospect of the Association paying off its bad debts. The great difficulty with members seemed to be the sale of their honey and to assist them the committee arranged to hold a honey fair on the 16th and 17th December last at Nottingham.

Messrs. Morris and Place kindly placed their mart at their disposal, and the amount of honey staged was a little over a ton, besides a small quantity of wax. The sales were, however, slow, only six or seven hundredweight being sold. At this fair the medals of the BBKA and their certificate were offered for competition. There were numerous competitors, including most of the leading beekeepers in the county.

The silver medal was won by Wm Silver, of Retford; the bronze medal by the Rev. RA McKee, of Farnsfield; and the certificate by Mr. GE Caparn, of Newton. The honey fair was a loss of £8 to the Association. The committee would be glad to arrange for lectures in different parts of the county if desired. Mr. EC Walton, of North Muskham, had resigned the office of expert in consequence of press of business. The balance-sheet for the year 1887 showed that the income had been £23 11S 2½d, and the expenditure £24 1s 6d.

Votes of thanks were then accorded to the retiring office bearers. The office-bearers for the ensuing year were then elected. The annual drawing for bee-keeping appliances, which had been presented by various manufacturers, was next proceeded with, and a vote of thanks to the chairman brought the meeting to a close.

BBJ 9 Feb 1888
For Sale. Vol. I. to X. British Bee Journal, bound in Publisher's covers, with Advertisements and extra blank pages. First-rate condition. Vol. II. minus Index, otherwise complete. Price £4 17s. Hd. Cheap. Rare opportunity. Address T. Lowth, Brant Broughton, Newark.

BBJ 1 Mar 1888
The prize schedule for bee-hives, honey, etc. at the Royal Agricultural Show, to be held at Nottingham in July next, is now ready, and may be obtained upon application to the Secretary, Mr. John Huckle, Kings Langley, Herts. Members and other bee-keepers who may be willing to forward specimens of bee-flora for decorating purposes are requested to communicate with the Secretary without delay. Arrangements will be made for the carriage of small parcels of such from those who may be willing to assist.

24 May 1888.
BBKA meeting. The secretary reported that upwards of 230 entries had been made for the bee department of the Royal Agricultural Show, to be held at Nottingham.

BBJ 24 May 1888
BBKA Quarterly Conference. Mr. T Lowth, of Brant Broughton, Newark, Notts, exhibited his 'Lowth's Unique Extractor.' This extractor is specially made for the purpose of extracting honey from the comb made in unshapely, unfinished, and

unsaleable one-pound sections. The machine will be found handy for extracting honey for exhibition purposes, the honey being easily ejected from delicate comb without a risk of damage in breakage and with a minimum of waste, the quality and brilliancy of the honey being preserved. Bee-keepers with small apiaries, who work on the section principle, will find this extractor desirable. The machine is provided with three extra cases for the purpose of holding comb honey to be extracted other than that of sections. The extractor can also be adapted for the reception of two-pound sections.

The illustration will furnish an explanation of its mode of working. The workmanship of Mr. Lowth's extractor was highly applauded, but it was thought that the work it could do could also be effected by the larger extractors.

Class 192 .—For useful inventions introduced since 1886 Certificate was given to Mr. Lowth, for what he calls his ' Unique ' extractor for sections.

BBJ 31 May 1888
What may be done in a Case of Emergency? If I shall not be trespassing on your valuable space I should like to say what experience has taught me last summer and this winter, as I think it may be interesting to some of your numerous readers. I live on the border of the remnant of Sherwood Forest, there being several apiaries in the village, my small one amongst the rest.

Last summer there was a swarm left my neighbour's apiary and took up its abode in an unused chimney. The owner not wishing to fetch them out, I asked if I might do so. Having got his consent I at once went to the occupier of the premises and asked his leave and it was given at once.

In examining the place, I found it was a chimney belonging to an out-building, the gable end towards the orchard, with a parapet wall about twelve feet high, flat top, running from the gable end.

I took with me an empty straw hive, also some scorched rag. The chimney was built up from the inside of the room, but fortunately there was a hole into the chimney I could use. I at once set fire to the rag, and with a thin rod of wire pushed it into the chimney, smoking the bees out at the top. I then placed a ladder on the wall and rested it against the gable end of the building. When at the top I found a flat stone on the top of the chimney, to the underside of which a previous stock of bees had built comb, but had died and left it unoccupied. The swarm had taken possession of the comb and had been working in it for several weeks.

When I lifted the stone I broke down some of the comb. But there was a good quantity left attached to the stone which I turned upside down and carried down on my head and laid it on the wall the same side up I had carried it down the ladder. I then pulled up some long grass and brushed the bees that had settled on the outside of the chimney into my skep, brought them down and placed them over the stone.

In an hour they had settled in the skep. In two hours I took a cloth and laid it on the grass in the orchard, carried down the stone and skep, placed them in the centre, tied the four corners above, putting a strong stick through. With the assistance of my son I removed them to my apiary, where I had an empty bar-frame hive with some comb worked out. I shook the bees out on the bars, having set them about two inches apart, covering quickly with an empty sack; they settled down in a short time.

The next day I put the bars the proper distance, the bees all the time being very busy. I examined the hive this morning, and find that they have gone through the winter well, and cover about six frames two inches deeper than standard size. Being the first bar-frame hive I possessed I still keep it, the bees always working fairly well in it.

If you think it will interest any of your numerous readers I shall be glad to say in another letter how I succeeded in transferring a stock the last week in November.

W Robinson, Mansfield Woodhouse.

BBJ 7 Jun 1888
New Beginner. Inquiries respecting the NBKA should be addressed to Mr FKH Fisher, Farnsfield.

NEP 10 July 1888
The Royal Show. The surroundings of the Royal Show Park this morning were exceedingly fine. The sun shone warmly, but the heat was softened by grateful breeze, which fluttered the tent and shook the canvas, but was no more distorting than that.

BBJ 19 Jul 1888
The annual exhibition of the BBKA, held in conjunction with the Royal Agricultural Society, took place at Wollaton Park, Nottingham, on the 10th to the 13th of this month. The weather was most unpropitious during the early part of the week, and the number of persons visiting the show grounds on the days on which the entrance-fees were 5s and 2s 6d was much less than on any former year. On the Wednesday the attendance was 9057.

The following are the list of awards to NBKA members:
Class 186—For the best and most complete frame-hive for general use—2, WB Baker, Muskham, Newark; 3, Turner & Sons, Radcliffe-on-Trent

BEEKEEPING IN VICTORIAN NOTTINGHAMSHIRE

Class 190—For the best feeder for quick autumn feeding, capable of holding at least five pounds of food at a time, price not to exceed 3s - 2, A. Simpson, Mansfield

Class 196—For the best twenty-four 1-lb. glass jars of run or extracted honey (approximate weight) - 1, A. Simpson

Class 197—For the best twelve 2-lb. glass jars of run or extracted honey (approximate weight) - 2, A. Simpson

Class 198—For the best exhibition of honey from one apiary, in quantity not less than one cwt. There were eleven entries, but few staged - 1, A. Simpson, 30s.

BBJ 19 Jul 1888

Comments on Royal Show. Were I a lady bee-keeper, I would make a new flag and give it to the BBKA ours looked decidedly shabby. The bee-tent, large enough for ordinary local shows, is altogether inadequate for the throngs that visit it on such occasions; moreover the one used last week at Nottingham sadly requires new netting. The internal arena is sufficiently large for the purpose, but the outer canvas circle requires to be larger and not so low down on people's heads. A crush-barrier is actually imperative on such days as last Thursday with its crowds of visitors.

NEP 26 Jul 1888

The members of the Sutton-in-Ashfield Horticultural Society held their fifteenth annual exhibition, yesterday, under more unfavourable conditions as regards the weather than have been experienced for several years.

The NBKA had decided to hold their annual show in connection with the exhibition, but the venture was by no means a success. The season has been a very disappointing one to keepers of bees, there being a great scarcity of honey, and the exhibits of yesterday were, in consequence of the adverse conditions, not up to the usual standard. There were in all seven classes, in two which—for bell glass comb honey and 24 small sections comb honey—there were no entries. Mr W Martin, the Hall, Wainfleet, appointed by the BBKA, acted as judge. There were four candidates for examination for expert certificate, Messrs. W Silver, Retford; A Simpson, Mansfield Woodhouse; J Rawson. sen., Selston and FHK Fisher, Farnsfield, all of whom came through successfully. Mr. Fisher, the secretary of NBKA, gave lectures in the bee tent during the afternoon and evening.

The awards were as follows:
Stock of bees of any race, be exhibited with their queen in an observatory hive—1st, A Simpson; 2, W Robinson, Woodhouse; 3, FHK Fisher
Best 12 1 lb sections of honey— 1st (silver medal of the BBKA), W Silver, Retford; 2nd, J Holmes, Hockerton
Best 1 lb. bottles of run or extracted honey—1st (bronze medal of the BBKA), A. Simpson; 2nd, J Rawson, sen.; 3rd, RW Pett, Widmerpool
Best run or extracted honey in white flint bottles, not exceeding 2 lb each with total weight of 24 lb—2nd, A Simpson
Bee-driving competition—1st, W Robinson: 2nd, RW Pett; 3rd, TS Rawson

The BBKA arranged to hold yearly examinations of candidates and to give first, second and third class certificates according to the proficiency of the candidate. The first examinations took place in South Kensington in 1883 and naturally the first experts were from the Home Counties.

BBJ 26 Jul 1888
Sltte ye grcatc Sfjotoe.
There was nothyng muche, wen alle ys fayde, tomake a note of atte ye Nottynghamc Showe, after that wone hath viewed ye divers novell devices vvyche aperc from tyme to tyme; ye fayme olde barrfrayme hyv, ye fayme olde feder, ye ydentical olde fayces of menne difplaing their wayrcs as were fenc of yore. Wonne varyetic there was, mayde by yc Committee of Brytifli Bekepers, wych atte ye firll fyt was barelie relyfhed by fomme, to wit:—ye displacement of yc olde veterane-in-chefe Baldewyn from ye be tent. We were tolde that ye chefe reafon of this furpryfe was that ye mynde of ye B.B.K.A. did lede them ynto chofyng a rite deft manne wych would do hys werke atte les code of ye peces than erltwyle, by reafone of hys livyng yn ye neborode of ye fhowe. Thys foundethe rite wele, but yt il affortyth wyth fome olde-fafhyoned notions of ufyng ye fervante wele wen he fervyth wele, in alle ye hete and aburdene of ye daie, yn fay re wether as well as fowle, but do not put ym afyde for eke doeyng nothyng amys. Yt behovcth wonne to fe ye Committee tak a lyke mynd unto wifdome ynto alle tbeyr doyngs. Agayn, 1 have onelie prayfc to fayc of Mailter Howarde, ye godeman that dyd hys werke yn ye be tent ryte nymblie, and hys talkyng was alle common fenfe. So wele dyd ye byilanderes lyk ye difcours of Maiiler Howarde that they prefled themfelves ynto ye faycred ynclofure ytfelf atte grete ry(k
of geten yltabed with be-ltyngs.

BEEKEEPING IN VICTORIAN NOTTINGHAMSHIRE

There was ate tymes barelie room eno' for ym to go onne with hys fhowe. I had almollc to pufh rudelie to get withyn hearyng of yc myftcrics of handlyng ye bes. There was a jouflyng or tryall of (kill of mills for rollynge, and alfo the werke of makyng bes-waxe ynto flietes yclept foundation, wcrewyth to fyl ye woden fraymes yn ye hyv; there were who gotte their fyngers burned wyth ye boylyng waxe, and there were who fmyled at yt.

There were fome fearfullie and wonderfullic mayde dyverfions yclept Extractors; fome were not eafily underftanded of ye people, for Darby and Joan would afkc yf ye thyng were for churnyng butere, and Joan would foe often turn yt round about, that wonne niakere dyd perforce tyc hys up agayn and agayn. I could not but remark on ye lykenefs of people to fome animals yn Regent's Parke, for whatever could be moved, lyfted, or turned round, had to be fenlelefllie werked jult as ye laft wonne werked yt. Maillers Abbotte, Blowc, Godeman (of Saynt Albans)

Neibor, Gryffyn (of ye dubbin), Meadowes, ye two Dickfons, were alle to the front with their wayres, and prefent themfelves. Maillers Raynor, Seager, Hookere, ye fon of Clure of ye Lathams, and W. B. Carre (ye devyfer of ye metaille endes) were alle there.

(' May my end be lyk hys,' W. B. C, myt be ye motto of ye next inventor, or elfe ' hys end was (i) pece.' Wonne of ye judges was a young man, but I wot he was an olde judge by ye waie he applyed hymfelftoye be furniture. Ye judges, Mr. Carre, Rev. G. Raynor, and MRW Martyn, were alle there, and yt ys to be hoped we fal here no more murmuryngs of ye pail by refbn of the paynes they toke to do what ys fayre among men. I elofelie fcanned their fayces and ufed my fmokere till I was wcarie of waytyng, fo how tired ye judges mult have bene! I trow ye judge is always wrong by ye nature of thynges, for he cannot pleafe alle altho' he tries. Hys office ys lyke unto ye drawing waterc out of a well with a feve. To ufe an old joke, 'ynfled of honey he getteth but whackes.' So we know ye very bell upright men are ye judges and awarde ye prizes to ye bell godes.

A brave and godelie fhowe of floures likelie for bes to cull ye neilar out of, was mayde by yc firm at Southall and got its mede of prayfe. Of ye glafs hyves there, ye feders and fraymes I will not prayfe wonne over another, and of new fangled notions there were alle too manic.

There were liyves biggeand hyves tal, Hyves brode and hyves imal, Hyves nYilovve and liyves depe, Ye routes flatte and ye roofes ftepe, But hyves to plefe ye alle.

X-TRACTOR.

BBJ 30 Jul 1888
North Notts. I have been thinking for some time that I would send you an Echo, but hoped it would be a better one than this. Up to the present date I have only taken

nineteen moderate 1-lb. sections from my eight hives; last year at this time I had taken 475 lbs. of section and extracted honey, as I work four of my hives for sections and four for extracted, so that you will see that we are no better off here than in other parts. But all my hives are crowded with bees and brood from top bar to bottom bar, but when I examined them a week ago, I do not think there would be an average of 2 lbs of honey per hive, although I have been feeding all through this month; but as I only give sufficient syrup to last them from day to day, they are unable to store any.

All hopes of any further surplus honey in this district are now at an end, but the lime-trees show an abundance of bloom, so that if we could only get a fortnight of fine weather for the bees to take advantage of them, it would be very satisfactory to all of us, as I am only a working man, and consequently find a considerable difference in being able to sell from £17 to £20 worth of honey to having to purchase 2 cwt. of sugar for feeding purposes.

It has been rain! rain! rain! and on Saturday last (28th) there was half-inch of rain registered, and the same person told me that the total for the month was four inches and two parts (2-100ths), and that it was the largest total for July for over twenty years past. On the 27th I drove two skeps for a friend, united the bees, and transferred them and brood into a bar-frame hive; but, although I have driven hundreds of skeps, I never saw so many bees in two skeps before, but only 4 or 6 lbs of honey, which evidently were remains of last year's stores. Natural swarms about here, even from skeps, have been few and far between, and I hear of a few having died of starvation already; and it is almost impossible to make skeppists believe that bees require feeding in June and July. Local Hon Secretary, NBKA.

BBJ 2 Aug 1888
Royal Show, Nottingham. Will you allow me space to reply to the remarks of 'Amateur Expert' on the above? He says, 'Were I a lady bee-keeper I would make a new flag.' If Mr. 'AE.' will kindly send the materials for a new flag to Mr. Huckle, I will undertake to find a lady who will make them up. *Nous verrons*.

'The bee-tent is inadequate.' 'The internal arena (magnificent phrase) is sufficiently large, but the outer circle requires to be larger.' It is thought that circular bee-tents are altogether a mistake; and it is difficult even for ' Mr. Howard's stentorian voice ' to be heard behind him in any sized tent, but if the outer circle were to be enlarged it would become quite impossible for half the people either to see or hear what is wanted in a semicircular tent, an amphitheatre in fact, where gentlemen who do not boast stentorian voices like Mr. Howard might be able to address the whole audience at once; and if ' AE.' will give a guinea towards it I shall be happy to do the same.

From what I know of the gentlemen who compose the Committee of the BBKA, I am sure they would be only too glad to act upon any workable suggestions; but when Mr. 'AE' recommends that ' more opportunities should be given to the appliance-manufacturers,' he is evidently not aware of the difficulties under which the Committee act. The show is not, as he states, 'under the management of the BBKA,' but very much under the management of the Royal Agricultural Society.

If the manufacturers would apply direct to the Secretary of the Royal, as manufacturers of other articles do, they would no doubt obtain as much space in the show-yard as they require. As to the 'hint from the Show at the Colonial,' 'AE.' possibly does not know that every pound of honey exhibited at Nottingham was sold; and he also, apparently, does not know much about the ways and rules of the Royal, if he imagines that such a form of sale as that conducted at the Hertfordshire Stall at the Colonial on Bank Holiday, or at the Canadian Exhibition, would be for one moment tolerated at the Royal Agricultural Show. Had he not omitted to bring that 'good parcel of flowers from my own garden,' he would have been in a better position to criticise those who did not respond to the request, or who did send the ' poor faded things.'

It is very easy to compare the appearance of honey, hives, &c, at the end of the week, with the spick-and-span newly painted machinery outside; but Mr. 'AE.' would be doing better service if he would kindly suggest how their freshness can be preserved after they have been handled, from curiosity or for instruction, by thousands of people. I can say from my own observation that the shed was constantly cleaned up, notwithstanding the crowds which were passing to and fro. As 'AE.' at all times claims the right of criticising others most freely, I trust he may be able to take my remarks in good part — One who was there.

BBJ 2 Aug 1888
Leicestershire BKA held its Seventh Annual Show on July 25th and 26th at Leicester in connexion with, and by the aid of, the Leicestershire Agricultural Society. Wednesday, 25th, was a wretchedly wet day, and the show grounds were a dank and desolate appearance; Thursday was fine and bright, and visitors crowded in from town and country.
Class VI. Best hive 10s 6d (three exhibits): 1st, Turner & Son [The hive here awarded third prize took first prize at Nottingham Royal Show.]
Class VII. — Best super for comb honey (three exhibits): Commended, Turner & Son. Special prize of 5s awarded to Mr. Lowth of Brant Broughton for section slinger.

Sheffield Daily Telegraph 24 Aug 1888
The Honey Harvest. As an instance of the entire failure of the honey harvest up to the present time this season, the local secretary of the NBKA at Retford states that his stock of bees this year has made 19 lb of honey as against 476 lb last year by the same strength. This it accounted for by the absence of white clover and the continual

wet, which washes away the little nectar there may be in the blooms. Further, when it has been fine the east winds have prevailed, and bees could not take advantage of it. As further proof of the failure, not one-tenth part of those who entered for the recent show at Nottingham were able to exhibit. Mr. Silver, the Retford local secretary, has again secured the first prize for the best 12 sections of honey — two years in succession.

Sheffield Daily Telegraph 31 Aug 1888
The exhibition of the Derbyshire BBKA was a good one, considering the unfavourable season, the collection of hives and appliances being extensive, and commanding much attention. The bee driving competition was of the usual attractive character, and it may be stated that Mr. AG Pugh, of Ambergate, found the queen bee in the short space of $1\frac{1}{2}$ minutes. The prize winners in this department included JW Rawson and BS Rawson. Both of Selston.

BBJ 20 Sep 1888
The Battle of the Bees. Doubtless there are many thousands of your ordinary readers who would be keenly interested in watching the progress of a real bee-battle — an attack by some, or all, the bees of one hive on the occupants of another hive, with the wicked intention of pilfering the honey which the industry of the hive attacked has gathered. Such an attack actually took place yesterday in my garden, and for the space of quite an hour I had an opportunity of observing the savageness and determination with which these intensely interesting creatures tight.

The first intimation I had of the disturbance was a very loud buzzing and humming in the neighbourhood of my smallest and weakest hive. On going near the hive I at once saw what the real state of affairs was. A detachment of bees from a neighbour's hive were storming my own with very great determination. Some were fighting in the air, and others were endeavouring to effect an entrance into the hive itself, but, so far as I could judge, were being gallantly repulsed, for many bees dropped dead at the entrance. Meanwhile I had thought of a plan to render the position of the defenders more secure. At the entrance to the hive I placed a piece of perforated zinc, with holes sufficiently large to admit of only one bee at a time to pass through.

This doubtless relieved them and those that had effected an entrance would have the warmest possible time of it. But reinforcements were continually arriving for the attacking army, and the position of my bees outside the hive was becoming more and more desperate. Eventually they were all killed or driven away; perhaps some regained the hive, but very many were dead and dying on the ground. Many of the enemy of course were amongst the number and the remainder took to their wings and disappeared. On going to the hive this morning I counted twenty-four dead bees being carried out by the survivors.

These were either my own bees who had died of their wounds, or, which is very probable, they were those of the enemy who had gamed an entrance. Some time must elapse before they will settle down to work again, for they are greatly excited and do not leave the immediate vicinity of the hive, Doubtless these splendid creatures are apprehensive of another attack on their storehouse and act accordingly. Their wonderful sagacity in selecting for attack the weakest hive and the unerring instinct which enables them to discriminate between friend and foe, place these insects amongst the most wonderful of Gods creatures, whether we consider their industry, their forethought, or their observance of the laws which regulate and govern the working of each individual hive.

HJ, Carlton, Worksop, Notts. Sept. 13 ('Daily News').

BBJ 3 Jan 1889

Let us look at the work of the BBKA. During the past year judges, examiners, experts, and bee-tents, have been sent to Nottinghamshire among other counties. It will be readily understood that these arrangements can only be effectively carried out at considerable expense and trouble. Four quarterly meetings have been held, at which useful discussions have taken place on various subjects appertaining to bee-culture; and in addition to the above the Bee Department of the Royal Agricultural Show at Nottingham must not be lost sight of.

BBJ 10 Jan 1889

We regret to hear that Mr. FKH Fisher is compelled to resign the post of Secretary to NBKA. We are pleased to announce that a successor has come forward to supply the vacant post.

NEP 28 Jan 1889

After such a terrible year as the last was for bee-keepers, it satisfactory to find from a report which was read at the meeting on Saturday that the Nottingham Association is doing well and that there is an increase in members as compared with last year. Now these members are just of the class which the Association most wishes to benefit, namely, cottagers. That a dozen, or even a score of hives of bees might be kept by rural cottagers who now have not even a single straw skep is evident enough to anyone passing through our villages, and that sensibly managed bees will not only pay, but pay well, is a second fact which experience has placed beyond challenge. Last season was altogether exceptional, the worst, we believe for twenty years or more, and those who contemplate bee-keeping ought not to be unduly discouraged by reports they may hear of there being no return of honey worth speaking of. The NBKA is doing good work, and will, we trust, go on and prosper.

BBJ 7 Feb 1889

The annual meeting of the NBKA was held on the 26th ult., at the People's Hall,

Heathcote-street, Nottingham. Ald. Manning, JP, presided, and among those also present were—the Rev. FH Slight, Woodborough; Messrs. RJ Turner, Radcliffe; WF Newman, Calverton; D Burnham, Flintham; G Hayes and AG Pugh, Beeston; J Pollard, Woodborough; Frank HK Fisher, Farnsfield (hon, sec), &c.

A letter had been received from the president of the Association (Viscount Newark, MP.) regretting his inability to be present.

The report of the committee, read by Mr. Fisher, expressed satisfaction at the improvement in the position of the Association, as shown by the increased number of members, there being now 83 as compared with 62 in 1887. The increase had been chiefly among cottagers, evidencing that the Association was carrying out the chief object of its existence—the benefit of the cottage bee-keeper.

The annual show of the Association was held at Sutton-in-Ashfield in connexion with the Sutton Horticultural Society's Show on July 23rd. Owing to the very bad season, however, there was but little competition. The silver and bronze medals of the BBKA were offered for competition, the former being won by Mr. Silver, of Retford, who was also the winner in 1887, and the latter by Mr. A Simpson, Mansfield Woodhouse. The judge appointed by the BBKA was Mr. W Martin, of Wainfleet, who, in addition to judging, examined four candidates for certificates as experts. It was gratifying to know that at this, the first examination in Notts, all the candidates, Messrs. Fisher, Rawson, Silver, and Simpson, passed.

The Secretary made a tour amongst the members as far as practicable in the autumn; and now there were four of the members holding certificates as experts it was thought the best thing that could be done would be to arrange for spring and autumn visits to all members who might wish it.

The committee regretted that Mr. Fisher, the hon. Secretary for the past two years, was unable, owing to other engagements, to retain the office. The balance-sheet showed receipts amounting to £25 10s $6^1/_2$d. After allowing for various items of expenditure, there remained a balance due to the treasurer of £2 16s $3^1/_2$d.

The Chairman congratulated the members upon the improved position of the Association. He had much pleasure in moving the adoption of the report and balance-sheet. The Rev. FG Slight seconded the resolution, which was agreed to.

On the proposal of Mr. D. Burnham thanks were accorded to the officers for their services during the past year.

Mr. Newman moved the election of Mr. Fisher as a Life Member of the Association, in consideration of his past efforts as Secretary. The resolution was unanimously adopted.

Viscount Newark, MP, was re-elected as President, with the Duke of Portland, Lord CC-Bentinck, Mrs. Robertson, Ald. Turney, JP, Ald. Manning, JP and Mr. Mansfield-Parkyns, as vice-presidents for the ensuing year. The committee having been appointed, Mr. AG Pugh was chosen as Honorary Secretary in place of Mr Fisher resigned, and Mr F Newman was elected as Treasurer. Mr. G Hayes, of Beeston, consented to act as Secretary for the Nottingham district.

The drawing for hives, etc. afterwards took place, the proceedings being brought to a close with a vote of thanks to the chairman.

Alderman John (later Sir John) Turney was born in 1839. He was Mayor from 1886 until 1888 and was also made Honorary Freeman of the city of Nottingham. He died in 1927.

John Arthur Charles James Cavendish-Bentinck, 6th Duke of Portland was born in Perth, Scotland in 1857. He held many significant political positions during his lifetime. He was married in 1889 in London. He lived in Welbeck Abbey and died in 1943. He is buried, as is traditional amongst this family, at St. Winifrid's church, Holbeck.

Mansfield Parkyns was born in Ruddington in 1823. He was the grand-grand grandson of wrestler Sir Thomas Parkyns of Bunny.

"The hand of a clever craftsman is seen in the poppyhead stalls, their shallow canopies carved with quatrefoils in tracery, and in the reading desks with symbols of the Evangelists; the craftsman was Mansfield Parkyns, who, after a life of great adventure, came to live at Woodborough Hall. He spent two years on the work with the help of two joiners, doing all the designing and most of the carving himself. It was a labour of love in memory of his wife, and it proved to be a memorial to himself, for we read that these choir stalls, the last work of his life, were designed, carved, and given on Christmas Eve 1893 by Mansfield Parkyns. He died on January 12, 1894, aged 70. He carved the screen across the tower arch, and designed the pulpit. He sleeps outside the church he helped to enrich."   *'Kings' England – Nottinghamshire', Arthur Mee 1935.*

NEP 3 Apr 1889
To Beekeepers For sale, cheap, two beehouses, hives, boards, eight feeders. Apply 86. Kirkewhite-street

BBJ July 1889
NBKA Shows to come.
    July 24.—Sutton-in-Ashfield.
    July 25.—Southwell.

July 30.—Farnsfield.
Aug. 5.—Beeston.
Aug. 15.—Woodborough and Epperstone.
Sept. 5.—Greasley and Selston.
Hon. Sec, AG Pugh, Mona-street, Beeston.

Sheffield Daily Telegraph 28 Jul 1889
The forty-seventh annual show in connection with the Chesterfield and East Derbyshire Agricultural Society was held yesterday on the Recreation Grounds Chesterfield. Unlike last year the weather was very favourable, rain falling frequently during tbe afternoon.
English bees 1st JW Rawson, Selston
Honey in sections 1st JW Rawson; 2nd. A Simpson, Mansfield Woodhouse

BBJ 1 Aug 1889
Lincolnshire BKA exhibition was held on July 24th and 25th in connexion with the Lincolnshire Agricultural Society's Show. The opening day was favoured by fine weather and as a consequence the attendance was large. Mr. Thomas Lowth exhibited a nicely-made little machine for extracting the honey from 1-lb sections.

BBJ 15 Aug 1889
Shows to come NBKA Sept. 5.—Greasley and Selston. Hon. Sec, AG Pugh, Mona-street, Beeston.

BBJ 22 Aug 1889
NBKA. This useful and thriving society, which has for its object the improvement of bee-culture, held its fifth annual exhibition in connexion with the local flower show at Beeston on Monday, August 5th. Some time ago the Society experienced one of those reverses of fortune which, unfortunately, are too often met with in the history of all institutions of a similar character; but since Mr. AG Pugh, of Beeston, has occupied the post of secretary, through his energetic exertions things have taken a turn for the better, and the Society is rapidly rising to the position which it should hold in the county. The new members up to date this year number nearly seventy and in consequence of this satisfactory increase the Association has been enabled to offer its members much greater advantages than it had done in the past. The Society has now a membership of 112. Local shows have been held this year in connexion with floral exhibitions at Sutton-in-Ashfield, Southwell, and Farnsfield, while others will be held on the occasion of the Woodborough and Epperstone Horticultural Society's Show on the 10th inst. and the Greasley, Selston, and Eastwood Show on the 5th of September. Prizes were offered yesterday for sections of comb honey, extracted honey, collections of bees, exhibits of beeswax, collections of hives and appliances and bee-driving. There were twenty-four exhibitors in the extracted honey class and

the collection was voted the best seen in the county this year and exceeded the show held in connexion with the Royal Agricultural Society in Wollaton Park. There were twelve exhibitors of sections of comb honey, and ten of beeswax, and nine observatory hives were shown. For the driving competition the competitors were also numerous, and the result was not known until late in the evening. Mr JM Hooker, St. John's, London, a late member of the committee of the BBKA, with which NBKA is affiliated, was judge.

Results:
Class 1.—Twelve 1-lb. sections of comb honey—1st.M. Lindley, Eastwood; 2nd. W. Measures, Upton; 3rd. J. Harvey, Skegby. Highly commended, Mrs. Wootton, Widmerpool
Class 2.—Twelve 1-lb. bottles of run or extracted honey —1st. T. Rawson, Selston; 2nd. J. Wilson, Langford Hall, Newark; 3rd. B. Rawson, Selston; Highly commended, Lord St. Vincent, Norton Disney; Mrs. Burnley, Blidworth; Commended, R. Turner & Son, Radcliffe
Class 4.—Exhibit of bees of any race, to be exhibited living with their queen in an observatory hive—1st. AG Pugh, Beeston; 2nd. JW Rawson, Selston; 3rd. T. Rawson
Class 5.—Exhibit of beeswax—1st. B. Rawson; 2nd. A. Simpson, Mansfield Woodhouse; 3rd. Lord St. Vincent
Special Class (open to all England)—Collection of hives and appliances.—1st. L. Turner & Son; 2nd. RW Pett, Greyfriar Gate, Nottingham
Class 6—Bee-driving—1st. AG Pugh; 2nd. G. Hayes, Beeston

BBJ 22 Aug 1889
Excited Bees. A correspondent asks for a reason for the following;
One afternoon last week one came rushing into the house saying a hen was being killed by the bees. She was found lying on the ground exhausted and with difficulty got away. Then nearly all the fowls of all ages were seen flying, jumping, and both frightened and hurt; indeed, too much frightened to be got under cover; but after a time it was suggested to use the syringe, which restored some quiet. The stinging went on for over an hour probably and the worst bees seemed to come from two hives, to which after a time the syringe was mainly directed. There were about fifty fowls and they are fed close to the hives. The hen first found seems likely to revive, but two chickens are dead. Of course the bee-keeper was a good deal stung on the hands. Some extracted combs had been put on the strawberry bed that morning, but there had been no previous attacks. Twenty-nine hens have since been sent away and so far there is quiet with the remainder.   Rev. T. Eagle Vicarage, Newark.
[We had, several years ago, a very similar instance of the irascibility of bees. The bees suddenly rushed out of their hive and fiercely attacked a hen with seven chickens, six of which died of the stings inflicted and the hen was in a very sad state for a week, but ultimately recovered. The cause we found to be that one of the frames, being too

narrow, had fallen from its place. The sudden noise alarmed the bees, they sallied forth and wreaked their vengeance on the innocent chickens. The probability is that something similar, either inside or outside the hives belonging to our correspondent, had caused an alarm to its inhabitants. Ed.]

BBJ 19 Sep 1889
Bee-Sting. Having had very much the same experience from a bee-sting as Mr. Thomas Fawcett, but having been more quickly cured, I thought a letter on the subject might not be out of place in the Journal. One day in the early summer I was manipulating the bees and I received one sting on the back of the head. I had the sting removed in about a minute and ammonia applied. I did not attempt to take it out myself, not wishing to lose a quantity of hair in the operation, and not being able to go round and look at the back of my own head. In a few seconds after the sting was removed I felt a sort of tickling sensation in my tongue and immediately afterwards in my hands also. Then my eyes felt queer and on looking into a looking-glass I found my eyelids swollen and the whites of my eyes bright red. I took some weak brandy and water, and went off to business. As I walked up the street matters did not improve, but grew worse; my lips swelled, my face became puffy and my throat felt rather 'choky;' so, having to pass the doctors house, I thought I had better call in. When the doctor saw me my face was a peculiar colour, and my lips nearly black, and from my whole appearance I might have had a snake-bite; my pulse was very weak, and a large rash appeared all over my body. The doctor gave me three drams of *sal volatile* and immediately afterwards two more drams of the same, and made me bathe the back of my neck with cold water and in less than half-an-hour from the time of being stung I was well again, though rather feeling the large dose of ammonia. I felt no effects whatever in a couple of hours or so. I had been stung often before, and I have been stung often since, but never, except on that occasion, have stings had such an effect upon me. I am no more proof against stings now than I was before; unfortunately they swell just as much as ever. My advice to bee-keepers who think they may be stung in a vein is to keep a supply of *sal volatile,* and if they begin to feel any of the symptoms I have described, take a dose at once; and in the event of then not having *sal volatile* handy, try a good dose of strong brandy and water.
       Crispin E. Smith, Vicar's Court, Southwell, Sept. 10th.

Crispin Edward Smith was born in Southwell in 1866. He was married in Plomesgate, Suffolk in 1895. Vicars' Court is a late 18th-century square of houses for the clergy, with the Dean's Residence at the eastern end of the Minster town.

BBJ 17 Oct 1889
Report from North Notts. Bees in this neighbourhood have all round done well during season just passed. Stocks which were strong in spring have, under good management, yielded an average of fifty or sixty pounds. Those not supered have

swarmed again and again so that, after losing the larger half of their stocks last winter and spring, the bee-keepers around are looking forward to another season to put them on their feet again. A ' Parson's ' only stock has given him 103 1-lb sections well filled, about 20 half filled, 8 fully-sealed bars, and have still more left in the body box than they can consume before next season.

The owner of five frame-hives (spring count) has taken nearly 4 cwt. of honey, mainly extracted, his largest take being 120 lbs extracted from one stock. A farm labourer with twenty stocks has taken just over 1000 lbs, besides increasing stock. A working joiner has taken nearly 60 lbs per hive from 18 stocks. These parties all live in separate parishes round Mansfield and all are past their 'prenticeship. The cottagers' skeps have yielded about an average of 15 lbs. My best one stock yield was 105 lbs extracted, all from sealed combs. This hive did not swarm. A stock from which the swarm flew away afterwards gathered a surplus of 120 lbs extracted.

Best lot of sections was 105 lbs from a swarm. The parent stock of this gave me about 20 lbs extracted. These stocks have all sufficient stores to carry them well into next spring. Our bees are all blacks, though some of them have a dash of Italian blood in them. My apiary is situated a good mile from the heather on what is left of the Forest. All readers of Robin Hood will remember Sherwood Forest. Most seasons we get some heather honey, some seasons a fair quantity, but this year we have to be content with half-filled sections and bars.

The last day the bees did any real work was July 8th; the clover was then at its best, but the weather broke and it has been more or less bad ever since. I hear of a few stocks which were in the midst of the heather having gathered a little surplus; it could not be much I fancy, for I have driven some large skeps with strong populations quite on the edge of the heather, and the honey in them was nearly all clover. There was empty comb enough to hold heather honey had it been there for gathering.

A Simpson.

BBJ 23 Oct 1889
The monthly meeting of the BBKA Committee was held at 105 Jermyn-street on the 23rd instant. The minutes of the last meeting were read and confirmed. The Secretary reported that the Notts. Association had contributed 10s to the Mansion House fund for conducting the case of objectors to the Railway Rates at the inquiry now being held by the Board of Trade.

BBJ 23 Jan 1890
The last was my third season in beekeeping and was a fairly successful one. Three hives yielded me an average of 50 lbs a-piece. Clover is our mainstay, helped considerably by fruit-honey and limes. There is no heather in the district.

This oppressively mild weather is playing havoc with the stores. Already I have had to resort to candy, although in autumn the brood chamber was well supplied. Aconites, snowdrops, and here and there a crocus, are in bloom. Strange to say, the bees do not work the *Chimonanthus fragrans*, which is flowering freely. Too sweet for them probably. WM Bird, Thrumpton, S. Notts.

W. Montague Bird, Solicitor, born in 1854 in Madras as a British subject.

The annual meeting of NBKA was held on February 1st 1890, in the People's Hall, Heathcote-street, Nottingham. Viscount St. Vincent took the chair. There was a very large attendance. The annual report was very satisfactory, no less than 68 new members having joined during the year, the total number being 121 now, as compared with 83 on December 31, 1888.

The balance-sheet showed total receipts £48 12s 7d, compared with £25 10s $6^{1}/_{2}$d for 1888. The expenses amounted to £4 2s 11d.

The season just passed had been a fairly good one in Notts, although very short, several members having obtained as much as 100 lbs. from one colony, while one had reached the remarkably good total of 206 lbs from one stock. Experts' examinations have been held in both second and third class, two members receiving third-class and four second-class certificates.

After the report and balance-sheet had been adopted, about sixty members sat down to a splendid meat tea.

Viscount St. Vincent was elected President in place of Lord Newark, MP, who became a Vice-President. Viscount St. Vincent in his address alluded to his own experience, which he described as not theoretical but thoroughly practical, having an apiary of twenty-six stocks, which he attends to himself, including making his own hives. Having made a humorous allusion to the 'pig-headedness' of some of the old school of skeppists, he regretted that the railway rates question would have a bad effect upon modern bee-keeping. The annual prize drawing then took place. Mr FHK Fisher of Farnsfield, read a paper on "The enemies of bees" which was followed by a discussion.

The second president of NBKA was Carnegie Robert Parker Jervis, 5th Viscount St. Vincent, who held office from 1st February 1890 until 2nd March 1907. In 1883 his estates included land in Norton Disney, 7 miles NE of Newark. At one time the pub in Norton Disney was called the "St. Vincent Arms."
He appears to have been popular with members, being heartily thanked for his services at every AGM though by 1897 his

apologies for absence are always recorded. Both he and his wife suffered ailments which required them to spend a long time in the warmth of the south of France.

The president was a source of great financial help to the association. Besides his annual subscription he gave prizes for the annual honey show, and every year he provided two hives for the prize draw held at the AGM.

BBJ 20 Feb 1890
Bee-Houses. Your correspondent will find such a shed as he suggests answer his purpose admirably, but he must not forget plenty of ventilation for the very hot weather and also, if possible, make provision for the sun to shine in on the hives once a day.

I have used a bee-shed for three years. I find it answers well especially as in it I can manipulate earlier in the morning, before I go to business, than I can do out of doors. The hives are set north-west, the only way I can have them, and they seem to do as well as the hives out of doors in a better position. My hives are fixed separately on shelves nailed to four light posts sunk into the ground, and quite independent of the shed floor-boards. I leave a space of half an inch between hive legs and shed floor-boards, which space is covered with strips of thick felt, such as is used by saddlers, tacked to the floor and to fit close up to the hive legs. There is thus no jarring of hives when walking about the shed, or opening and shutting doors.

The tunnels for entrances are put through the wall of the building on the same principle. I have the hive floor-board fast to the hive and projecting about nine inches in front. I form a square tunnel, covered with either glass or board. I find board the best and have it fastened down. I fix slides to regulate the entrance close up to the hive and which can be handled easily inside the shed. I then cut a square opening in the shed wall, half an inch larger than outside of the hive tunnel. I then nail strips of thick felt around this opening on the inner wall to fit close up to the tunnel. This quite prevents outside bees from getting into the shed, and also quite hinders any vibration of the building from affecting the hive.

You can slide the hives in and out of the opening easily. You can fix a porch outside over the entrance. My building has three windows, and out of the bottom of each bottom pane is cut a strip three-eighths of an inch deep for inside bees to escape through. I used to have it at the top of the window, but I find the bottom is the best. There is one defect in this plan: sometimes outside bees will get in this way, and especially wasps.
If you have only one window and if that is a revolving one, you will easily release insiders, and prevent outsiders from getting in. I find they do not eat so much food in

winter as those out of doors, and are much quieter to handle. I have a loft above in which I keep empty hives, crates, &c.   Sampson White, Bleasby, Southwell.

W BROWN, Muskham, NEWARK
Hives, sound and good from 5s
Foundation, Sections, Smokers, Knives, Veils
Extractors from 7s. Metal ends and all appliances
Catalogues free

Advertisement BBJ 29 May 1890

Sheffield Daily Telegraph 11 Jun 1890
Notts. Agricultural Society. Arrangements have been made with the NBKA to hold an annual exhibition in connection with the societies meeting.

BBJ 19 Jun 1890
July 16-17. NBKA in connexion with the Notts. Agricultural Society's Show at Wollaton Park, Nottingham. Entries close June 2lst. Schedules, &c., AG Pugh, Hon. Sec, NBKA, 4o Mona-street, Beeston.

NEP 21 Jun 1890
Notts. Agricultural Show. The fact that the entries are due to close to-day reminds us that are once more within measurable distance of the annual show in connection with the Notts. Agricultural Society. The experiment tried last year of holding it in the Cattle Market—although in all respects successful caused protests from all classes of visitors because the nature of the surroundings. Hence the society could not but take the matter into consideration, to whether a more suitable site could not be secured. The question once raised, attention turned naturally to Wollaton Park, the scene of the successful Nottingham Royal, and it is satisfactory to note that, with Lord Middleton's permission, this admirable site has been fixed upon for this year's exhibition, to be held on July 16th and 17th.

Practical illustrations of manipulating with live bees, showing the best methods of driving, capturing the queen, transferring combs from straw sleeps bar frame hives, uniting stocks, etc, during the exhibition made by able masters. Arrangements will be made by which visitors may view with safety the mysteries the hive and witness the perfect command the scientific apiarian has over his bees, and to show how needless is the cruel practice of killing bees to obtain honey even under the straw hive system of bee-keeping.

BBJ 26 Jun 1890
H. (Nottingham) The queen sent is a remarkably fine one, but quite young and

unfertilised. We prefer to leave brood in the section till it hatches out. Cutting it away is a 'messy' job, and the section of comb never looks well afterwards, even if rebuilt and filled. To prevent it use an excluder below.

BBJ 7 Jul 1890
JG Peet (Nottingham). There is a trace of Carniolan blood about the queen sent, but she has apparently been mated with a black drone, and her progeny are practically 'blacks.'

JWG Kirk (Eagle Vicarage, Newark). Queen sent is a young one. By some mischance the old queen has got lost when the first swarm returned to the hive. The swarm which issued a week or so later might very possibly have more than one young queen with it. Sometimes several queens accompany the swarm under such conditions.

NEP 16 Jul 1890
In brilliantly fine weather the annual show of the Notts. Agricultural Society opened at ten o'clock this morning in Wollaton Park. The day after St. Swithin, the sun shone fiercely—an agreeable change after the cold and wet weather of the past fortnight. Special trains are being run from all parts of the county, and there is every prospect of big attendance in the course of the day. The scene in the park is reminiscent of that, though of course upon a diminutive scale, when the Royal Agricultural Society of England pitched their tents in Lord Middleton's beautiful demesne. A prominent feature in former exhibitions—the poultry show—is absent this year, but there is a big display of bees and honey, which somewhat compensates for this.

Bees, Honey, etc. It was a source great disappointment to the executive and members of the NBKA that upon this occasion, when they had such fine opportunity of displaying the growth of bee culture in the county and explaining the techniques of the art, that the season has handicapped them so very severely. Wet days, cold nights, and little sunshine all tell against the beekeeper, and these conditions having been the rule during the past 10 days, have given him but a small chance. And it happened that, although the Beekeepers' Association tent was by far the most important side show of the agricultural exhibition proper, and notwithstanding the numerical strength of the entries obtained, the display of hived bees and honey was not large as had been expected. It may be mentioned, showing that bee culture is not upon the decline, that 50 new members have been elected this year to NBKA, and Viscount St. Vincent, the president, does much to encourage the interesting and profitable pursuit.

Mr. RW Pett, of Greyfriar-gate, Nottingham, had a most diversified collection of hives and appliances, including a new frame hive and an elaborate observatory hive, which took first prize in Class 1, and Mr. EC Walton, who was second, showed quite a novelty in the shape of a live wasps' nest, with the busy, spiteful, little creatures hard work building up. Mr. Pett was also successful taking first prize for the most complete frame hive with a new design, and again he was to the fore with the best specimen bees of any race exhibited in observatory hive, his queen bee being particularly fine and large.

The honey shown was of good quality, though the scarcity of the commodity so far this season considerably reduced the amount shown. A new feature was an exhibit of granulated honey, intended to demonstrate to the sceptical that granulated honey was absolutely honest and unadulterated, whereas some of the fine flaked honeys may treated with glucose or some other ingredient, and yet appear pure. Mr Wm. Silver, Retford, sent a very fine sample. The BBKA silver medal, together with Lord St. Vincent's special prize for the best jar of run extracted honey, was secured by Mr. T Wilson, of Newark, and similar prizes for the best 12 sections of comb honey went to Mr. FHK Fisher, of Farnsfield, who is an enthusiast in the work of the association. The exhibition of bees was chiefly remarkable for a particularly beautiful specimen from Lord St. Vincent, which was very easily first. The contents of the beekeepers' tent were inspected with considerable curiosity, and doubtless tomorrow there will be a further big attendance to witness the bee-driving competition. Mr. P Scattergood, jun., of Stapleford, and Mr. M Lindley, Beauvale, were courteous stewards in the tent to-day.

Class 1 — For the best collection of hives and appliances to consist of the following articles (exhibits in this class to be staged by the exhibitor):- 1 frame hive, priced at 15s; 1 ditto, priced 10s 6d. (Note - these hives must be fitted with arrangements for storifying); 1 observatory hive, 1 pair section crates fitted with sections, 1 extractor, 1 slow stimulating feeder, 1 rapid feeder, 1 smoker or other instrument for quieting bees, 1 veil, 1 swarm box for travelling, capable being used as a nucleus hive, 1 travelling crate for comb honey, 5 other distinct articles—not specified—at the discretion of the exhibitor—1, RW Pett. 12, Greyfriar-gate, Nottingham
Class 2 —For the best and most complete frame hive—1. RW Pett
Class 3 —For the best specimen of bees (any rare) exhibited with the queen in an observatory hive—1, RW Pett, 2, Wm K. Kirk, Eagle Vicarage, Newark-on-Trent; 3, Arthur G. Pugh, Mona-street, Beeston; 4, J. Clarke, Loscoe, Codnor
Class 4 -For the best exhibit of run or extracted honey, jars not exceeding 1-lb each to approximate 12 lbs— 1, J. Wilson, The Gardens, Langford Hall, Newark-on-Trent; 2, BS Rawson, Selston, Alfreton; 3, JW Rawson, Town Green, Selston, Alfreton; 4, TS Rawson. Town Green, Selston, Alfreton; 5, FC Piggin, Hucknall Torkard
Class 5—For the best twelve sections of comb honey, the gross weight to be approximately 12 lbs —1, Frank HK Fisher, Farnsfield; 2, Matthew Lindley, Newthorpe; 3. W. Measures, Upton 4, Lord St. Vincent, Norton Disney, Newark-on-Trent; 5, JW Rawson

BEEKEEPING IN VICTORIAN NOTTINGHAMSHIRE

Class 6 —For the best exhibit of six 1-lb bottles of granulated honey—1, William Silver, Bridge-gate, Retford; 2, J. Wilson; 3. RW Pett

Class 7.—For the best exhibit of beeswax, not less than 2 lbs —1, Lord St. Vincent; 2, J. Wilson; 3, BS Rawson

Frederick C Piggin was born in Hucknall in 1879. The Piggin family were supporters of the New Methodist cause with the Rev John Piggin as their leader. He was a butcher in the town.

Matthew Lindley was shown in the 1881 Census as a carpenter.

BBJ 31 Jul 1890

The annual show of NBKA was held in connexion with the Notts. Agricultural Society at Wollaton Park, Nottingham, on July 16th and 17th, and but for the unfavourable season an exceedingly good and large show would have resulted. It was a source of disappointment to the executive of the Bee Association that the season handicapped them so very severely, and tended to reduce the quantity and quality of the display. The NBKA is one of the most active associations now working; fifty new members have joined this year and the very active Hon. Sec, Mr. Pugh, is deserving of all praise for the energy with which he pushes on with the work.

Unlike most Bee Associations, the NBKA is fortunate in having for president a nobleman (Lord St. Vincent) who is himself a bee-keeper and besides taking a special interest in the bee department, is an exhibitor and prize-winner. The arrangements for the show were excellent, and reflected credit on the Association and its indefatigable Hon. Sec; in fact, with fine weather, a capital attendance—especially on the second day—and the bee-tent well patronised, no drawback was experienced beyond what was entirely attributable to the poor season. The entries in the bee and honey and hive classes numbered 116.

List of Prizes.

Appliances

Best collection of hives and appliances—1, RW Pett; 2, EC Walton

Best and most complete frame hive, price not to exceed 10s 6d—1, RW Pett; 2, EC Walton

Bees.— Best specimen of bees (any race), exhibited with the queen in an observatory hive — 1, RW Pett: 2, ME Kirk; 3, Arthur G Pugh; 4, J. Clarke

Honey, etc

For the best twelve one-pound jars run or extracted honey—1, J. Wilson; 2, BS Rawson; 3, JW Rawson; 4, TS Rawson; FC Piggin

For the best twelve sections of comb honey — 1, Frank HK Fisher; 2, Matthew Lindley; 3, W. Measures; 4, Lord St. Vincent; 5, JW Rawson

Best six one-pound bottles of granulated honey—1, William Silver; 2, J. Wilson; 3,

RW Pett

Best exhibit of beeswax, not less than two pounds—1, Lord St. Vincent; 2, Thomas J. Gosling; 3, BS Rawson

Bee Driving Competition.—1, J. White, Eastwood; 2, AG Pugh, Beeston; 3, G Hayes, Beeston; 4, RJ Glew, Clifton

Robert J Glew was born in Wroot, Lincolnshire in 1856. He is recorded in the 1881 Census as a corn miller assistant but by this time he was a corn miller.

BBJ 26 Sep 1890

Stung by a Queen. I send you particulars of a new experience to me, and possibly to most other beekeepers. We are told by many writers that a queen has never been known to sting any person, but my case stands thus:

When driving a skep the other day, I placed the queen in my mouth, as is my usual habit and had scarcely done so when I was astonished to receive a very sharp sting on the lip. I need hardly say the queen was quickly returned into the hive. I have watched her with no little curiosity to see if she was harmed by stinging me, but to-day (16th) she is alive and laying freely.   RW Pett, Greyfriar Gate, Nottingham.

[Only in very rare instances does a queen-bee use her sting when being handled, and in your case it would probably be but an involuntary prick caused by a little roughness or pressure while placing her in your mouth. They do sometimes bite (or what may be called a bite) when held too tightly between the lips, but it is to our mind a rather foolish 'habit' to place a queen in the mouth, except when really necessary for the purpose of freeing the hands. Eds.]

BBJ 9 Oct 1890

Successful candidate for BBKA third-class experts' certificates, who have passed during the year 1890: John B White, Eastwood, Notts.

BBJ 16 October 1890

J. Glew (Newark) The heather sent is the true honey-producing variety. Bear in mind, however, that it must be grown by the acre or the 100 acres before the place can be called a bee-pasture.

BBJ 29 Jan 1891

NBKA. The annual meeting of the above very active Association takes place at the People's Hall, Nottingham, on Saturday, the 7th of February. The President of the Association, Viscount St. Vincent (himself a successful beekeeper), presides on the occasion and will be supported by the Mayor of Nottingham (Samuel E Sands). At the conclusion of the business portion of the proceedings, which commence at 3.30pm, the members and their friends partake of a substantial 'meat tea.' A social evening follows, when the annual prize drawing takes place. Then follows a lecture on 'Bee-keeping' by Mr. John H. Howard. Altogether the proceedings are so arranged as to make up a very enjoyable evening, and reflect much credit on the gentlemen concerned in making the arrangements.

NEP 7 Feb 1891

The annual meeting of NBKA was held at the People's Hall this afternoon, when there was a good attendance. Viscount St. Vincent presided, and among those present were the Mayor of Nottingham (Mr. SE Sands), and Messrs. P. Scattergood, Stapleford; W. Lindley, Eastwood; J. Rawson, Westwood; J. White, Eastwood; Geo Hayes, Beeston; AG Pugh, Beeston (hon. secretary); J. Bowley, Stapleford, FC Piggin, Hucknall; TJ Gosling, Arnold, J Farnsworth. Eastwood; T. Simmons, Lane, Pinxton; Watts, West Bridgford; Holmes, Edwalton; Roe, West Bridgford and Derry, Nottingham.

The Hon. Secretary read the annual report, from which appeared that the year 1890 was not such a successful one for individual beekeepers as was at one time expected. They would, however, be pleased learn that NBKA had made progress. They had 121 subscribers during the past year, but of that number three were donors, who did not renew their donations this year, four members left the country, 10 resigned, and unfortunately, by the death of Viscountess Ossington they had lost one of their chief subscribers. They had, however, 64 new members enrolled, that being an increase of 46, and their membership now reached 167. The membership had just doubled itself during the past two years. (Applause.)

The balance-sheet, although still showing adverse account, was more satisfactory than last year. During the season shows had been held in connection with the Notts. Agricultural Society, and local horticultural societies, Beeston, Southwell, Sutton-in-Ashfield, and Moor Green, and to all those societies they were deeply indebted.

Their thanks were also due and tendered to the president Viscount St. Vincent and Mrs. Chambers for their liberality in increasing the prize fund. Mr. Scattergood moved the adoption of the report and in doing so said every member had shown a great desire to put the association on a sound footing, and the committee had given members every possible help to advance the best interests of the beekeepers of the county. The committee had to a large extent succeeded in carrying out its desire. Mr. Simmons seconded the proposition, which was unanimously adopted.

The balance sheet for the year ending December 31st, 1890, showed that the receipts had been: Subscriptions, £31 3s 6d; donations, £3 4s income from other sources, £28 9s 2d total, £62 16s 8d. There is balance due the treasurer £1 4s $1^{1}/_{2}$d. The meeting then adopted the accounts.

Viscount St. Vincent, referring to the society's subscription to the BBKA, said he did not object to that guinea being paid, but he did think that the BBKA did not do anything for them in the way they might do. The British Society had great show at Doncaster on the 20th of June, but that would be no benefit to them as it would be impossible for them to exhibit. Mr. Howard, a member of the British Society, said the guinea subscribed to the British Society was money well spent. For that the association received a silver medal, bronze medal, and certificate for exhibits which were shown.

Mr. Pugh then moved that a vote of thanks be accorded the Mayor. This was seconded by Mr. Wootton and Mr. Hayes and carried. The Mayor, responding, said he was pleased to have had the opportunity of listening to the report of their work, and the very worthy object they had view. Very few people took interest in beekeeping, but he thought that was in a measure due to their not understanding the insects. They, at all events, very little understood the instinct of bees. Without bees they would very soon short of some of their finest flowers. He was delighted to hear such a favourable report, and would, indeed, pleased to become one of their vice-presidents.

The meeting then adjourned for tea.

Charlotte (née Cavendish-Bentinck), Viscountess Ossington (1806-1889), Daughter of 4th Duke of Portland; wife of 1st Viscount Ossington. When she was twenty years of age, and when her father refused to hear of the suit of John Evelyn Denison, Lady Charlotte expressed her intention of eloping with Mr. Denison, and at the prospect of indirectly creating a sensation in high life the Farmer Duke relented. Lady Charlotte's marriage was her first triumph. Her next was when her husband rose to be Speaker of the House of Commons In 1857 and she herself one of the most important personages at the Court of Queen Victoria. She was to become the third President of NBKA.

Early in 1881, Charlotte, Viscountess Ossington, proposed to present the town of Newark with a new cafe or "Coffee Tavern". It was to be built close to the River Trent, opposite the Castle and cattle market, on land purchased from the Handley family. It was opened in 1882 in the presence of Lady Ossington. The purpose of the coffee house was "to promote the cause of temperance therein" and hopefully lure farmers away from the town's public houses. The building is still there in 2018 and is being used as a restaurant.

BEEKEEPING IN VICTORIAN NOTTINGHAMSHIRE

At that time the cattle market was just on the other side of the river Trent from the Ossington but is today (2018) further to the north on the old North Road.

BBJ 26 Feb 1891
The annual meeting of NBKA was held at Nottingham on February 1st 1891—the President, Lord St. Vincent, presiding. A large company, including the Mayor of Nottingham, assembled, among those present being many influential beekeepers of the county, including several ladies.

According to the annual report, read by the Hon. Secretary, Mr. AG Pugh, NBKA continues to make progress, showing a net increase of forty-six members since 1889. The number last year was 167. The principal show of the year was that held in connexion with the Notts. Agricultural Society at Wollaton Park. Minor shows have also been held in conjunction with the local horticultural societies at Beeston, Southwell, Sutton-in-Ashfield, Arnold, and Moorgreen, at all of which considerable interest was aroused.

The executive of the Association are making strenuous efforts to render bee-keeping popular, and are arranging to assist members with advice and help through the agency of certified experts, etc.

In responding to a vote of thanks, the Mayor expressed the pleasure it gave him to hear so favourable a report of their position and trusted Lord St. Vincent would honour them by presiding over their meetings for many years to come. For himself, he gladly accepted their invitation to become one of the Vice-Presidents and hoped to be present at their next meeting. After the meeting in the afternoon about 110 persons sat down to a substantial tea. Subsequently another meeting was held, Lord St. Vincent presiding.

The following gentlemen were re-elected to act on the Committee:
Messrs. FHK Fisher, M. Lindley, SW Marriott, WF Newman, H. Price, A. Warner, and S. White; Messrs. A. Simpson, Rawson, sen., Baguley, Watts, JE Phillips, and Thompson were elected in the place of the retiring members. Mr. P. Scattergood was appointed Auditor; Mr. AG Pugh, Hon. Treasurer and Secretary; Mr. JE White, Assistant Secretary. The Vice-Presidents, with the addition of the Mayor, were re-elected.

In the course of the evening Mr. John Howard, of Holme, Peterborough, gave an interesting lecture on bee-keeping, which was heartily appreciated.

BBJ 11 Jun 1891
Conference of Bee-keepers at Doncaster. Mr. AG Pugh, Hon. Sec, NBKA, writes:

'I am sorry to see so little notice is being taken in your correspondence columns and elsewhere of Mr. Coxon's suggestion that a friendly gathering should be convened at the Royal Show at Doncaster. I am sure many of us who propose attending would have been delighted to have known at what hour, date and place we should be able to meet a few of the fraternity, and I trust it is not too late to have it inserted in your Journal. I trust it may be on Thursday, the 25th, as that will be likely to be the most popular day.'

BBJ 18 June 1891
Conference of Bee-keepers at Doncaster - We have just received the following note: 'Sir,—I fully endorse the remarks of Mr. Pugh in last week's issue respecting the friendly gathering of beekeepers at the Royal Show. I shall try to get over, merely for the apiarian department, and should be most happy to meet "a few of the fraternity," as I consider this golden opportunity should not be lost. To me, a chat with others in the craft is one of the greatest luxuries connected with beekeeping. I hope something definite has been arranged.   O. Wootton, Draycott, Derby, June 15th.'

Our correspondent overlooks the difficulties attending his proposal. First, there is no room or building available for such a meeting within the show-ground, and second, it is quite certain that a meeting held outside would not be well attended. We, therefore, can do no more than say that both Editors of the BBJ will be present on the 22nd, 23rd, and 24th, and Mr. Cowan, as steward of the Bee Department, will of course remain until the close.

BBJ 18 Jun 1891
July 15th, 16th. Notts. Agricultural Society at Nottingham. Bees, honey, and appliances. Entries close June 20th. For schedules, &c, AG Pugh, Hon. Sec, NBKA, 49 Mona-street, Beeston.

BBJ 2 Jul 1891
Experts' Certificates. I should like to have the views of other experts upon the certificate given to those who pass the BBKA Expert Examinations. I consider it of a most paltry character, and, whereas many of us would be pleased to have a nice, bright, artistic certificate, framed and hung up in our homes, the present one is usually consigned to oblivion. Considering that a better certificate is given for honey, and that all first and second-class competitors pay the parent Association a fair examination fee, I think we ought to have something better.
                                                        AG Pugh, Second-class Expert.

BBJ 2 Jul 1891
RWP (Notts.) Comb is broody—not very bad, as bees are hatching from the cells while here; but the disease is unmistakably there.

BEEKEEPING IN VICTORIAN NOTTINGHAMSHIRE

NEP 15 Jul 1891

Notts. Agricultural Show. In the large bee tent there was a multifarious collection of articles appertaining to apiculture, there being on exhibition all the latest contrivances for extracting honey, bee driving, etc. There were also hives of every shape and size, while diagrams illustrating the various parts of bees microscopically enlarged were hung in places. "A Hallamshire beekeeper" had sent for exhibition, though not for competition, an observatory hive of "Punic" bees (*Apis niges*), and there several other observatory hives full of bees, the hives, of course, competing for the prize offered in their class. There was a large quantity of pure honey and beeswax on view, and exhibitions of bee driving and manipulations with live bees were given during the day in a special tent set apart for the purpose.

BBJ 23 Jul 1891

Forcing Hives for Clover Harvest

1. Would cork-dust do in lieu of chaff for winter covers? I mean, would it do to fill the bags with to put over the quilt?
2. Would it be advisable to put any insect-repellent in the cork-dust placed in the double walls, to prevent the cork from becoming the home of woodlice, &c.?
3. In our district the bees never seem to do much till the white clover begins, and then they seem able to gather almost any amount of honey, as the district is one of the best for white clover I ever saw. Within 500 yards of my house there are dozens of fields quite white with it at the present time, and mine is the only apiary within three miles. I have noticed that in yield of honey, late hives, that is, those that are weak in April, gather as much honey and are as strong when the honey season begins, as those that are then relatively much stronger; hence it has occurred to me that it is comparatively useless to stimulate in March and April, so as to get the hives crowded with bees in May, some time before there is a real flow of honey. Is this a sound opinion? — Pedagogue, Notts.

Reply. 1. Cork-dust is superior to chaff for the purpose.
2. Anything likely to keep away obnoxious insects is useful, but we should not specially urge the need for it.
3. It is sound reasoning to say that there is less need for forcing on bees in early spring in order to prepare for a late harvest, such as clover. Six weeks is the orthodox time for building up a stock, and if the bees will make themselves strong in time for clover without the forcing process, there is no need for adopting it.

BBJ 30 Jul 1891

NBKA. This flourishing Association held its annual county show in connexion with the Notts. Agricultural Society, in Wollaton Park, on July 15th and 16th. The weather was everything that could be desired, and there was a very large attendance, the bee and honey tent coming in for a large share of attention, while the bee-driving contests created a considerable amount of interest.

Class 1, for collections of hives and appliances (three entries)—Only two exhibitors put in an appearance, but both made an excellent show, Mr. Walton's collection being especially good, nearly every article required in modern beekeeping being shown, and all proving the exhibitor to be fully competent to keep up the reputation he has already acquired. 1st prize, EC Walton, Muskham, Newark; 2nd, RW Pett, Greyfriars-gate, Nottingham

Class 2, for the most complete and inexpensive hive for cottagers use (five entries)— All the hives exhibited in this class were of sterling value, and the judges had some difficulty in awarding prizes, which were given as follows 2nd, EC Walton; 3rd, RW Pett

Class 3, for best specimen of bees in observatory hive (nine entries) — This class provoked a keen competition, and was a source of much interest to the visitors. Mr. Walton had an easy first with the hive he exhibited at the Royal Show at Doncaster. The second prize was awarded to Mr. Pett for a hive of his own construction, containing two frames, and worked on the principle of a toilet looking-glass, enabling the spectator to turn the frames in the best position for getting full advantage of the light. 1st prize, EC Walton; 2nd, RW Pett; 3rd, C. Clarke, Loscoe Grange, Codnor; highly commended, J. Clarke, Loscoe Grange, Codnor

Class 4, best twelve bottles of extracted honey —Out of twenty-two entries in this class only fourteen exhibits were staged, owing to the backward season. These were, however, nearly all of exceptionally good colour and flavour, and were only in a few cases rather devoid of consistency. Mr. Wilson, who takes the silver medal in this class, was also the winner of this much-coveted prize last year. 1st prize, J. Wilson, Langford, Newark; 2nd, EC Walton; 3rd, Viscount St. Vincent, Norton Disney, Newark; 4th, FC Piggin, Hucknall Torkard; 5th, FHK Fisher, Farnsfield, Southwell

Class 5, best twelve sections — The exceptionally backward weather experienced in Notts. was especially shown in this class, for although seventeen entries had been made, only four exhibits were staged, each taking a prize. The first prize—which, with other first prizes, had been given by the Association President was won, together with the bronze medal, by his own sections, much to the satisfaction of the members. He, however, further demonstrated his liberality and good wishes for the cause of beekeeping by informing the Secretary that the money he won in this and other classes was to be retained towards paying the incidental expenses of the show. 1st prize, Viscount St. Vincent; 2nd, W. Measures, Upton, Southwell 3rd, EC Walton; 4th, RW Pett

Class 6, for best six bottles of granulated honey—Only two exhibitors entered honey in this class, and as neither were good, the first prize was withheld ; 2nd, FC Piggin

Class 8, for best sample of beeswax (seven entries)—The President scored another first in this class with a very superior exhibit. 1st, Viscount St. Vincent; 2nd, J. Wilson; 3rd, HJ Raven, West Bridgford

Class 9, bee-driving — Seven competitors entered for this class, and great interest was taken in it by the public. 1st, GH Merrick, Hucknall Torkard; 2nd, AG Pugh,

BEEKEEPING IN VICTORIAN NOTTINGHAMSHIRE

Beeston; 3rd, FC Piggin; 4th, JB White, New Eastwood
Special class, for best six pounds of honey produced within four miles of Nottingham marketplace — This being a new class did not meet with the patronage it is hoped it will do another year. 1st prize, AG Pugh; 2nd, Beecroft, Wilford

Altogether, the show of 1891 may be considered a very successful one, the only drawback being that the fine weather caused several members who would have been present as competitors to stay at home in the hayfields.
[An esteemed Notts. correspondent sends us the above report. Eds.]

It is interesting to note that neither of the prizewinners in the special class lived within four miles of Nottingham market place!

NEP 7 Aug 1891
Moorgreen Show. There was a large tent set aside for, amongst other things, exhibiting observatory hives. The honey was judged by Mr RJ Turner of Radcliffe

BBJ 6 Aug 1891
Lincolnshire BKA Show. Best 24lbs extracted honey – 2nd, EC Walton, Muskham, Newark

BBJ 13 Aug 1891
Yorkshire BKA Show. Complete frame hive – 2nd, EC Walton, Muskham, Newark

BBJ 20 Aug 1891
NBKA, besides its annual show in connexion with the Nottinghamshire Agricultural Society, holds several other exhibitions with local horticultural and agricultural societies. One of these took place on Bank Holiday, August 3rd, in conjunction with the Beeston Horticultural Society. The morning opened fine, but during the afternoon rain came down in torrents. Notwithstanding this, a fair number of visitors attended the show, the bee and honey-tent coming in for a large share of attention.

Class 1. Best twelve sections of comb honey — This class does not appear to be worked to any great extent by the bee-keepers in Notts. and only four exhibits were staged, and though each received a prize, there was a keen competition before the awards were made as follows:—First prize, Mrs. Hind; second, Mr. K. Merrick; third, Mr. Measures; fourth, Mr. Pett
Class 2. Best twelve 1-lb. jars of extracted honey — Fourteen exhibits were staged in this class, all of exceptionally good colour and flavour. A few of the exhibits lacked consistency, owing, no doubt, to the honey being extracted before it was sealed. Mrs. Hind again took first prize; second, Mr. J Wilson (winner of the silver medal at the county show); third, Mr. Raven, of Bridgford; fourth, Mr. JR Swift, Arnold

Class 3. Best observatory hive stocked with bees and their queen — Four hives were shown in this class, and were a source of much interest to the visitors, who appeared very anxious to catch a glimpse of the 'queen.' Mr. J. Clarke, of Loscoe, was first with a splendid exhibit stocked with Carniolans; Messrs. RW Pett and FC Piggin being placed equal second.

Class 4. Best exhibit of beeswax. — First, Mr. J. Wilson; second, Mr. HJ Raven

Taken altogether, the show was in every way a success, and the stewards, Messrs. Bowler and Baguley, had bestowed every care in staging the exhibits to the best advantage.

BBJ 20 Aug 1891
To Be Sold Two Hives of Bees (Swarms). Price and particulars of WH Glossop, Forest House, Babworth, near Retford.

BBJ 3 Sep 1891
AM (Nottingham) The sample sent is cane sugar.

BBJ 10 Sep 1891
For Sale. Eight Strong Stocks of Bees in Bar-frame Hives, cheap, in consequence of Removal. For price and particulars apply to Smurthwaite, Sutton, Retford.

BBJ 19 Nov 1891
STP (Nottingham).
Candy Feeding. A cake of soft candy may be laid over the feedhole in enamel-cloth quilt without disturbing the latter, provided the candy is covered with a box-lid, or some such thing, to keep in the bees. The amount of food you name (fourteen pounds) will no doubt suffice till March next, but it is best to err on the safe side.

BBJ 17 Dec 1891
Leaving on Surplus Chambers. Width of Entrance during Winter.
1. I have three hives some miles from home, and not having a convenient place to store supers, &c, have left them on the hives, i.e., two with worked-out shallow frames and one with sections. The two with frames in the supers have excluders on, the sections have not. There is a fair amount of stores and bees in the hives, but no honey in the supers, and they are well supplied with quilts (over supers). Would the bees winter well under these conditions? — and,
2, How wide would you advise entrances to be? AGP (AG Pugh?), Notts.
Reply. 1. So long as there is no honey in the surplus chambers no harm will arise beyond the inevitable loss of heat to the cluster of bees below. We have, however, not seldom wintered bees under exactly similar conditions to those named, and had excellent results.

2. Much depends on circumstances, i.e., the form of hive used; method of wintering; exposed or sheltered situation, &c. Our own hives have outer cases, and they are located in a rather exposed position as regards cold winds, consequently entrances to the outside are only left open about three-quarters of an inch wide during the winter months. But it must not be forgotten that the combs of each stock are raised three inches above the floor-board; also that the entrance to the hive itself—from the space between it and the outer case—is fourteen inches wide. Hives without outer cases, and where no space below combs is allowed in winter, should have entrances left at least six inches wide, except in the face of cold, biting winds.

BBJ 24 Dec 1891
Sad Accident to a well-Known Beekeeper. We have received from Lord St. Vincent, President of NBKA the particulars given below. The first account sent by his Lordship stated that Mr. Pugh had died from the effects of his injuries the same night, but, though the report was apparently well founded, it happily proved incorrect. Just before going to press we learn from Lord St. Vincent that Mr. Pugh still lives and we are sure that all will rejoice to know of even this small consolation.

The following is the letter referred to:
'It is with feelings of the deepest regret that I write to inform you of a very sad accident which happened to Mr. Arthur G. Pugh, who, while following his usual employment, on the 16th inst., seems to have stepped out of the way of one train into the front of another. The train passed over both legs, and severely injured his head. He was conveyed by the same train to Leeds and taken to the infirmary, where, from later advices, I learn that, though very severely injured, he still lives and is progressing as favourably as can be expected.

'Mr. Pugh, who for a long time has been Hon. Secretary of the NBKA, will be sadly missed by all Notts. beekeepers who have always found him most kind, courteous, considerate, and helpful in every difficulty. The harder the work that he had to perform, the more cheerful he seemed to be. The Association has lost its right hand, at least for a long time, and all members of it a friend whom they will find it hard, if not impossible, to replace.
    I am, &c, St. Vincent, President NBKA, December 19th, 1891.'

BBJ 14 Jan 1892
The Accident to Mr. AG Pugh. Doubtless many of your readers will be pleased to hear an account of the progress made by our esteemed Hon. Sec, Mr. Pugh, so well known amongst bee-keepers. I called at the infirmary in Leeds, and found our friend progressing so quickly that we soon hope to have him amongst us again. Mr. Pugh has had a large portion of each foot amputated, and also received injuries to face and legs; his life was despaired of, but, thanks to his splendid physical condition and

temperate habits, recovery seems now assured. His mental faculties are as good as ever, and he reads all our bee-literature, and is arranging a report, &c, for annual meeting, and we may well hope that he may yet be enabled to resume his old post of Hon. Sec. of NBKA, the duties of which are now performed by Mr. White, assistant secretary. RJ Glew Newark-on-Trent.

NEP 28 Jan 1892
Hucknall Beekeepers' Conference. After tea last night a conference of local beekeepers was held in the Coffee Tavern Hall, when the subjects of bee-driving, treatment of disease, doubling hives, queen catching, etc. were debated. It was resolved get the Notts. Association Bee Tent at the next Hucknall Show, and liberal promises of donations for the prize list were made by those present. Messrs. FC Piggin, Wiggett, Lodge, and Cartledge took a prominent part in the proceedings.

The annual meeting of NBKA was held on February 27th, 1892 at the People's Hall, Heathcote-street, Nottingham. The President (Viscount St. Vincent) sent a telegram expressing his regret that he was not well enough to attend the meeting and the chair was taken by Mr. FHK Fisher (Farnsfield). Mr. Hill and Mr. Wootton, of the Derbyshire BKA, were amongst those present.

The balance-sheet showed that the total receipts for the year amounted to £71 4s 5d and that, after meeting current expenses, the adverse balance of £6 9s due to the Treasurer at the outset of the year had been reduced to £1 12s 7d. Thirty pounds (within sixpence) was distributed in prizes at the six shows held during the season at Wollaton, Moorgreen, Beeston, Southwell, Hucknall, and Arnold. The accounts and report were approved and passed.

Viscount St. Vincent was re-elected President. The Duke of Portland, Lord Newark, MP, the Mayor of Nottingham (Richard Fitzhugh), Ald. Manning, JP, Mrs. Hind (Papplewick), and Mrs. Chambers (Alfreton), were elected Vice-Presidents, and the following Committee was appointed: Messrs. Marriott, Warner, Baguley, Rawson, Lindley, Forbes, Watts, White, Fisher, Simpson (Mansfield Woodhouse), Simmons, and Poxon. Mr. Scattergood was re-elected Auditor and in proposing the re-election of Mr. Arthur G. Pugh to fill the joint office of Treasurer and Secretary, suggested that a telegram be sent informing him of his re-appointment and sympathising with him in the accident he had sustained, which prevented his attendance. This was unanimously approved and agreed to.

Mr. John White was reappointed Assistant Secretary. It was stated that in order to ascertain the amount of honey gathered during the season the President had offered prizes for the best results. Mr. John Howard delivered a lecture on beekeeping, in the course of which he gave practical hints and replied to a number of questions. A discussion followed, and the usual drawing for prizes concluded the business.

Since the above meeting we are glad to learn that Mr. Pugh has so far recovered as to return to his home, and that, although still weak, he hopes to be able to attend to all indoor work connected with the secretaryship of the Society.

Mansfield, February 6th, 1892. I began bee-keeping in May, 1890, but the bees died the following winter. I think they were not strong enough to withstand the cold. I got another stock last September. The owner said they had plenty of stores for winter, but I gave them a large cake of candy. I examined them this afternoon—that is, took off the quilts—and found all the honey gone and the candy also. They just seem to have finished the candy. I gave them another cake, and also changed the calico cover, which was slightly damp. The bees seem fairly strong and lively.

Loxley J Meggitt.

Loxley J Meggitt was born in Conisborough, Yorkshire in 1874. In the 1881 Census he was shown as an analytical chemist.

NEP 11 Apr 1892
Bingham. On Saturday night the east of the town was made the scene of some smart pranks. Soon after midnight gates were taken their hinges, several yards of walls partially broken down, and gardens trodden over, and there is little doubt that some beehives would have been overturned had not the owner arrived upon the scene. Not long since other walls in the same locality were similarly treated.

NEP 30 Jun 1892
Technical Instruction in Bee-keeping. We are pleased to learn that the Notts. County Council have awarded the sum of £40 to NBKA for the purpose of furthering technical education in bee-keeping in the county. It is proposed to expend the sum of £20 in providing an outfit of bee-appliances and the remaining £20 for lectures.

BBJ 6 Sep 1892
NBKA annual show at Moorgreen—bees, hives, and honey. Honey classes confined to Notts. only. Entries close August 20th. For schedule apply to AG Pugh, Secretary, Mona-street, Beeston. Other shows in connexion with NBKA will be held at Hucknall Torkard, July 26th; Southwell, July 28th; Beeston, August 1st; Thorneywood Chase, August 11th; and Mapperley, August 1st and 2nd.

BBJ 17 Sep 1892
NBKA Annual Show. The yearly display of honey, etc. of this important Association was given in connexion with the Greasley Agricultural Show, and this, also, is an aggregation of the Greasley, Selston, and Eastwood Agricultural and Horticultural Societies. We suppose this is on the basis that "United we stand; divided we fall," for

stand they do. In the heart of estates belonging to Earl Cowper, it is gratifying to find such an excellent understanding between all classes, and such a capital exhibition of farmers' and gardeners' produce. Most satisfactory of all was it to find such an association of enthusiastic beekeepers, and it has never been our lot to see more friendly intimacy and goodwill, in spite of the keenest rivalry, almost reaching to disagreement in its earnestness. This is as it should be amongst us—we can agree to differ, and differ to agree. The Notts. bee-keepers seem brethren, even if, like other brethren, they do not all think alike.

All our readers who call to mind the severe accident to Mr. AG Pugh, of NBKA, last year, will be glad to learn that he has so far recovered as to resume his avocation as an active hon. secretary. The judge was Mr. RAH Grimshaw, and Mr. CN White was engaged as lecturer, whose services were utilised by the judge in assisting him to award the prizes for honey, the entries being so numerous that it was more than one man's work to give such a capital show its due.

After the awards were given by the judge, six candidates for third-class experts' certificates were examined by him. Nothing remains to be said but that the staging, space, and exhibits were worthy of a Royal Show, with the exception of the class for appliances (makers, please note).

Prize List.
Collection of appliances—1st prize, RW Pett, Nottingham
Best hive—1st, WP Meadows, Syston; 2nd, AW Pett
Observatory hive—1st, P. Scattergood, jun., Stapleford; 2nd, W. Brooks, New Eastwood; 3rd, GH Merrick, Hucknall
Best twelve bottles of honey—1st, AG Pugh, Beeston; 2nd, J. Wilson, Langford; 3rd, H. Merryweather, jun., Southwell; 4th, Mrs. Hind, Papplewick; 5th, FHK Fisher, Farnsfield
Best twelve sections — Viscount St. Vincent, Norton Disney; 2nd, FHK Fisher; 3rd, Mrs. Hind
Best granulated honey—1st, RW Pett; 2nd, J. Wilson; h. com., Viscount St. Vincent
Best frame of honey - 1st, Mrs. Hind.
Best beeswax—1st, J. Wilson; 2nd, Viscount St. Vincent; com., M. Lindley, Newthorpe
Bee-driving—1st, TS Elliott, Southwell; 3rd, AG Pugh
Amateurs' sections—1st, J. Finn, Basford
Amateurs' run honey—1st, J. Finn; 2nd, JJ Taylor, Nuttall

Dr. Thomas Stokoe Elliott was born in Southwell in 1872. He qualified as a BBKA expert – second class – in 1894. His beekeeping activities will be found throughout this work.

He served in the RAMC 1914-1922. The citation for his Military Cross in 1918 reads "Capt. (A/Maj.) Thomas Stokoe Elliot, RAMC. For conspicuous gallantry and devotion to duty. This officer was unremitting in superintending the dressing of the wounded during four days' fighting. He was in charge of three advanced dressing stations, each of which was in turn destroyed by enemy shell fire, but he managed to evacuate all the wounded. Officers, personnel and patients were all encouraged by his cool example." In the Honours List for 1918 he was awarded a Bar to his MC for services in France and Flanders. He died in Salisbury, Southern Rhodesia in 1922.

Henry Merryweather was born in Carlton-on-Trent in 1839 and died there in 1932. In 1856 he came across a type of apple he did not recognise. He found out that the tree from which they came was located in Southwell. Henry worked up a stock of trees from seeds of the parent tree. They are now known all over the world as Bramley apples.

It was at this show that W Herrod was examined for the BBKA third class certificate but did not succeed.

NEP 2 Aug 1892
Southwell Horticultural and Cottage Gardening Society. Exhibition of Roses, Plants, Fruits, Vegetables, Bees, Honey, Butter, Eggs, etc. Thursday, July 28th, 1892. Notts. County Council Demonstration in Butter-making and Beekeeping. Old Robin Hood Band—2s Performers. Southwell Grammar School Sports. Dancing in the Evening Illuminations. Balloons. Special Train from Nottingham, leaving 1.50 pm., returning at 9.30 pm., fare 1s

Yesterday the Beeston Horticultural Society held their annual show of flowers, fruits, and vegetables, in a field on the Wollaton-road. This year, last, the committee were unfortunate in having bad weather for their show, as rain fell heavily about half past two o'clock, and probably deterred several people attending. When, however, the weather cleared, as it did, a more enjoyable afternoon was passed. The exhibits were numerous and of fine quality. There was an exhibition of hives, bees, honey, and appliances used in modern bee culture, this being under the auspices of NBKA. For 1 lb sections of comb honey Mrs. Hind of Papplewick, and Mr. Measures were equal, and Mr. Wright was third. Mrs. Hind was also first for 12 1-lb bottles of run extracted honey, with Mr. Wilson second, Mrs. Hill third, and Mr. Wright fourth. The other awards were— Collection of bees any race, to be exhibited with their queen in observatory hive—1st, Mr. Wootton: 2nd, Mr. Hill; 3rd, Mr. Merrick. Beeswax—1st, Mr. Wilson; 2nd, Mr. Wright.

BBJ 18 Aug 1892
Lincolnshire BKA. This annual show and gala took place yesterday in Boultham Park. The weather was dull and cloudy during the morning, but the proceedings were favoured with fine weather until about five o'clock, when a shower fell. The show maybe, however, described a success, and new features were introduced, a rabbit and cavy show, and bee and honey competition, drawing a fair number of entries.

Honey Judge was Mr. FJ Cribb, Gainsborough.

For the best stock of English bees with queen, exhibited in an observatory hive, showing combs on both sides, Mr. Godson took first prize. For the best 12 one pound sections of comb honey prizes were awarded to Mr. T. Lowth, Lord St. Vincent, and Mr. R. Godson. For the best sample of run or extracted honey the first prize was won by Mr. Taylor, the second by Mr. Lowth, and the third Mr. G. Taylor. For the beat samples of bees wax, not under 3 lb weight, prizes were awarded to Lord St. Vincent, Mrs. Emerson, and Mr. Emerson. In the cottager class, for best glass comb honey not under 10lb in weight, WH Spengler and G. Taylor were successful. In the bee tent illustrated lectures on the manipulation of bees were given by Mr Cribb.

FJ Cribb lived in Retford until 1892 and worked at R Hornsby and Son as chief engineer and works manager. He then moved to Morton, Gainsborough. He became secretary of the Gainsborough district BKA in 1893 and three years later was appointed expert to the Lincolnshire BKA. He travelled extensively in that county giving lectures on bees sometimes accompanied by Dr Percy Sharp.

Welbeck Show Catalogue 1892
The Welbeck Tenants' Agricultural Society was established before the end of the 1880s. Its main business was to host annual shows in which tenants of the Welbeck estate would compete in various classes. The main show was held annually in July or August. It included prizes for horses, cattle, sheep and pigs; and also special prizes for garden produce, poultry, honey, and for long service.

NEP 25 Oct 1892
Technical Instruction. Lectures and demonstrations on beekeeping have been given under the auspices of the NBKA at four centres—Hucknall Torkard, Southwell, Gotham, and Moor Green—and had been well attended and appreciated.

BBJ 19 Jan 1893
Standard Honey Bottles. When I introduced this subject in your issue of October 6th, 1892, little did I think that it would assume such proportions, and that we should have such a long and interesting correspondence, or that it would be noticed and taken up by our leading bee-men, much less be the subject of the editors' notes. I have been looking through the whole of the letters which have appeared, and I think, on the whole, the matter has been taken up in a reasonable manner. Of course, we cannot all think alike, or act alike, either in this or any other matter and I cannot see the force of the editors' remarks in "Useful Hints," December 8th, where he says, "The bee-keeper who gets a good price for his honey should be honest enough to give the buyer a full sixteen-ounce bottle for his money, while the less fortunate one, who deals with shop-keepers, is no less honest in putting up his produce in 'reputed

BEEKEEPING IN VICTORIAN NOTTINGHAMSHIRE

pound ' jars and selling it at per dozen jars."

"That is the rub," and perhaps this is the crux of the whole question: if bottles and honey are sold at so much a dozen and not per pound, then it may be all right; but I think it is unfair to the public (who in this district, at any rate, look upon a bottle of honey as a pound) to receive anything short, and just for a little extra profit to only give fourteen ounces. And, with "Bee-Kay," (p. 482) I think we should maintain the credit of bee-keepers at large, and be above suspicion even in this small matter.

One of your correspondents, AP Wilmot (p. 509), hopes that total abstinence does not often lead to such ignorance as not to know the contents of a wine bottle. However much he may regret my ignorance on this point, I will frankly and freely say I do not, and I think if he had had my experience he would say the same and not be at all anxious to know the contents, either measure or otherwise, of a wine bottle. I am more than pleased with the discussion and hope that it will have well served its purpose and that the BBKA will at no distant date take the matter up and, as our friend "X-Tractor" from the "Hut" says in your issue of January 12th, 1893, fix on an all-round best bottle for the bee-keeper.

Now, sirs, for a personal matter. A number of my friends here who know my *nom de plume* say I ought to sign my own name and not write under an assumed one. I do not think so, at any rate for the present, but I will say that I am an official of NBKA and my home and apiary as the crow flies are not more than half a mile from the Hemlock Stone.

BBJ 14 Jan 1893.
[The above subject has now been so thoroughly discussed that its future consideration may, we think, be left to the Committee of the BBKA, if that body decides to make any recommendation as to a standard honey jar. After next week we therefore propose to close the correspondence. In the meantime, readers who have not yet contributed their views on the subject and desire to do so, will please take note of this. Eds.]

BBJ 26 Jan 1893
Should "Hemlock Stone" or any "seventeen-ounces-to-the- pound" bee-men come my way, I may say there is a four-and-a-half-gallon cask of mead in my cellar, made in 1891, a glass of which would quickly soothe any of their qualms of conscience and also, by putting them into a "fair" frame of mind, prove to them that justice is not to be found in either extreme. Nemo.

Standard Honey Jars. It seems to me that this correspondence has gone rather wide of the mark. It is simply a matter of detail between purchaser and seller what kind or size of receptacle honey should be sold in and I take it the BBKA have nothing to do with this. What is required, however—and, moreover, what I know several of your correspondents alluded to—is that all honey staged for exhibition at our honey shows shall be shown in exactly similar bottles and it is thought by some to be the duty of the BBKA Committee to decide upon a "standard bottle" for this purpose, leaving each to choose the bottle that suits him best for ordinary sales.

Arthur G Pugh, Hon. Sec. NBKA.

NEP 17 Feb 1893

Lecture at Orston. An interesting lecture, illustrated by lantern views, on "Bees and their management" was given in the National Schoolroom last night by Mr. Scattergood, of NBKA, under the auspices of the Notts. County Council Technical Education Committee. The lecturer was introduced by the Rev. TW Swann, vicar, and a considerable number of parishioners and others were in attendance.

Rev Thomas Wilson Swann, MA, vicar of Orston-cum-Thoroton. Born in 1840 in Bedale, Yorkshire. For twenty-six years he was assistant diocesan inspector, Southwell and member of Bingham Rural District Council. He died in 1910.

BBJ 23 Feb 1893

The annual general meeting of the members of NBKA was held on February 11th at the People's Hall, Nottingham—Viscount St. Vincent, President, in the chair. There was a large attendance, including Mr. FHK Fisher, Mr. P Scattergood, Jun., Rev. JS Wright, Messrs. J Marriott, C. Redshaw, Warner, Gosling, White, Young, Wootton, Hayes, Baguley, Wilson, Meadows, AG Pugh (Hon. Sec), and others.

Viscount St. Vincent, in opening the proceedings, said he was afraid he could not altogether congratulate them on the excellence of the honey season, which had not been quite as good as in some previous years. After all, however, the year had not been such a bad one for some of the members, at any rate. They had also had a grant given to them by the Nottingham County Council. His Lordship concluded his remarks by relating a few humorous experiences of his own at bee-shows.

The annual report stated that there had been an increase of membership during the year. The number of members on the books on December 31st, 1891, was 184, and on December 31st, 1892, 195, or a net increase of eleven members for the year. The Committee regretted that financially they could not report as favourably as they had hoped to do, there being a slight balance still due to the treasurer. The report also referred to the grant by the County Council and to the engagement of lecturers, who had addressed meetings at various places in the district. The report and balance-sheet were agreed to.

Mr. Wootton, after moving a vote of thanks to the retiring Committee, proposed that the following be the Committee for the ensuing year:—Messrs. J Baguley, FHK Fisher, M. Lindley, SW Marriott, W. Poxon, JT Faulconbridge, C Forbes, J Rawson, Sen., J Finn, T. Simmons, S. White, G. Wood, and the District Hon. Secretaries. The motion was unanimously adopted.

The following were also elected: President, Viscount St. Vincent; Vice-Presidents, the Duke of Portland, Lord Newark, MP, Councillor Sands, JP, Mr. JE Ellis, MP, Ald. Manning, JP, Rev Watkin Homfray, MA, Mrs. JEF Chambers, and Mrs. Hind; Auditor, Mr. P Scattergood, jun.; Hon. Treasurer and Secretary, Mr. AG Pugh.

A hearty vote of thanks to Viscount St. Vincent, for his services as President, was unanimously adopted.

After the members had taken tea together, an interesting ceremony took place in the presentation of a purse of money to the Hon. Secretary, Mr. Pugh, who, about a year ago, sustained serious injuries on the railway. Viscount St. Vincent, in making the presentation, said the members of the Society felt they would like to display their sympathy with Mr. Pugh, and they had taken this form of showing that sympathy. It was a great pleasure to him to have the privilege of making this presentation and he hoped Mr. Pugh would he long spared to act as their Secretary. Mr. Pugh said the presentation came quite as a surprise to him. He was deeply indebted to the members for their kindness, and he hoped that he should always be worthy of their confidence and that he would be able for many years to continue his connexion with the Society.

The distribution of prizes for best honey returns having taken place, Mr RAH Grimshaw, of Horsforth, near Leeds, delivered a lecture on "Foul Brood: What it is, and how to prevent it," and we hope in the near future to give a full report of the lecture in our pages. A short discussion on the lecture afterwards took place, followed by a vote of thanks to Mr. Grimshaw for his interesting and instructive lecture.

The meeting terminated with the usual prize drawing, Lord St. Vincent giving a "Wells" and one other hive, and the Association a hive and an extractor, &c.

Rev JS Wright was headteacher at Southwell school by 1908 having succeeded his father. In 1902 he subscribed £5 to aid in the repairing of the school.

Rev. Watkin Homfray was rector of Retford St Michael the Archangel's church in 1866. In 1899 he supervised the lengthening of the chancel. In 1899 a carved oak chancel screen was installed in his memory.

Neighbour's "Wells" Hive

The "Wells" hive was a two-queen system conceived by Mr. Wells of Kent in about 1892. He seems to be the only man who ever succeeded in working this type of hive. W Herrod commented, "It is significant that those who purchased this type of hive (or in the case of NBKA won one in the prize drawing at the AGM) very quickly adapted them into chicken coops or dog kennels."

NG 3 Mar 1893
County Council Lectures. Last night a lecture on "Beekeeping", profusely illustrated with lantern views, was delivered by Mr P Scattergood at Hucknall Coffee Tavern Hall.

BBJ 9 Mar 1893
Wanted. A Foreman Joiner who understands Bee Appliances and Greenhouse Work preferred. Must be sober and have a good character. Cottage on premises. Apply, stating wages, to EC Walton, Muskham, Newark.

BBJ 16 Mar 1893
On Thursday night, February 16th, a highly interesting lecture on "Bees and their Management" was given in the National Schoolroom, Orston, by Mr. Scattergood, in connexion with the Technical Education Committee of the Notts. County Council. The lecturer, who was introduced by the Rev. TW Swann, vicar, explained that the County Council conferred a grant on the Nottinghamshire Beekeepers' Association, who in their turn had laid on him the pleasing obligation of delivering this and similar lectures on apiculture. Mr. Turner, Radcliffe-on-Trent, assisted with the lantern, by which the

lecturer's remarks were capitally illustrated. A description of the queen, the drones, and workers was graphically given, and their duties clearly set forth. The lecturer mentioned that of the varieties of bees, his preference lay with the British black bee, and expressed his conviction that bee-culture, with patience and perseverance, would be found to pay, and would be a suitable addition to the occupations of cottagers, artisans, small farmers, and others. At the close of the meeting, questions were asked, and received satisfactory answers from Mr. Scattergood.

BBJ 11 May 1893
For Sale. 5 doz 16-oz. Screw-capped Bottles of good Granulated Honey, 1892. Cash offers. Address Elliot, Rectory, Southwell, Notts.

BBJ 25 May 1893
King Bee (Notts.) Flower sent yields honey, but is never grown in sufficient quantity to render its value of any importance as a honey plant.

BBJ 25 May 1893
Working Foreman wanted in a Bee and Poultry Appliance Factory. For full particulars apply to EC Walton, Muskham, Newark.

BBJ 22 Jun 1893
August 7th and 8th. NBKA Annual County Show at Mapperley in connexion with the Porchester Horticultural Society. Entries close July 31st. AG Pugh, Secretary, 1 Mona-street, Beeston.

BBJ 29 Jun 1893
Two Queens from Large Stocks, 2s 6d each. Address Harry J. Raven, Park Avenue, West Bridgford.

NEP 7 Aug 1893
The Porchester Allotment Gardens Floral and Horticultural Society, at their two days' exhibition, which was opened this afternoon in field of the Duke of Cambridge Hotel, Woodborough Road. A number of samples of honey were staged and in this connection it is worthy to note that a similar exhibition held at last year's show has been the means of inducing some eight or ten allotment holders to commence the occupation of bee-keeper.

BBJ 10 Aug 1893
NBKA Shows at Hucknall-Torkard on July 25th; Southwell, July 27th; and Arnold, July 31st.

That the present season has been one favourable for the bee-keepers of Notts. both for quantity and quality—and especially the latter —is amply proved by the exhibits at the above named shows. It is hardly too much to say that, as far as the extracted honey was concerned, almost every entry was of very high excellence, both as regards quality and style of get-up and we think that here is shown unmistakably the result of the years of painstaking work and instruction of NBKA and its energetic secretary. Though a very busy man, Mr. Pugh finds time to arrange a larger number of shows than are held in any other single county, and, through the help of the local secretaries, all the details are perfect, and each show goes off without the least hitch or confusion.

At Hucknall-Torkard there were no less than twenty entries for extracted and seven for honey in sections. The first prize for extracted fell to some rather dark honey, just delicately flavoured with heather—this being gathered on the Derbyshire border of the county. The other prizes went to local samples; these, in most cases, being very good in colour especially, though most had a peculiar flavour, the bees having evidently found some unusual source of supply in this exceptionally hot and dry season.

The awards were as follows:
For best 12 sections comb honey—1st, Mr. Wiggett, Hucknall; 2nd, Mr. Pett, Nottingham; 3rd, Mrs. Hind, Papplewick
For best 12 1-lb. bottles run or extracted honey—1st, Mr. Wootton, Draycott; 2nd, Dr. TS Elliott, Southwell; 3rd, Mr TH Frommings, Radcliffe; 4th, Mr. AG Pugh, Beeston
For best specimen of bees and queen in a unicomb hive—1st, Mr. Pett; 3rd, Mr. Wiggett; 4th, Mr. WH Cartledge, Hucknall
For best 6 jars of run or extracted honey (local)—1st, Mr. Wiggett; 2nd, Mr. WH Cartledge; 3rd, Mr. WH Cartledge
For best 2 shallow frames of honey—1st, WH Cartledge; 2nd, Mr. Wiggett; 3rd, Mrs. Hind

At Southwell the judge had a difficult task, a finer or more even lot of extracted honey being very rarely seen. There were twenty-one entries for extracted, and twelve for comb. The first-prize honey at Hucknall here again secured the premier award. The other prizes were awarded to samples that would have been prize-takers anywhere, being for flavour, ripeness, and colour, about perfect. Comb honey in sections was, all round, better than at Hucknall, one exhibit, in glass sections, being especially well filled and of fine flavour; this secured second prize. The comb honey, in bell-glasses, was quite a large class, and though fair, calls for no special remark, as only one sample was of really fine flavour.

The awards were as under:
For best 4 2-lb sections—1st, Mr. Marshall, Norwell

BEEKEEPING IN VICTORIAN NOTTINGHAMSHIRE

For best glass super of comb honey—1st, Rev. HL Williams. Bleasby; 2nd, Mr. Geo. Wood, Oxton; 3rd, Mrs. C. White, Newark
For best 6 2-lb sections comb honey—1st, H. Merryweather, Southwell; 2nd, Mr. AJ Mortimer, Oxton; 3rd, Mr. Measures, Upton
For the best 6 1-lb jars run or extracted honey— 2nd, Mr. Lee, Southwell; 3rd, Mr. Mortimer
For best unicomb hive with bees and queen —1st, H. Merryweather; 2nd, Mr. Mortimer

At Arnold the extracted honey was again of high excellence, but the comb honey hardly up to the mark of that at the previous shows. Awards as below:
Best 6 sections of comb honey—1st, H. Merryweather, Southwell; 2nd, Mr. Pett, Nottingham; equal 3rd, Mr. Wootton, Draycott, and Mr. Brooks, Eastwood
Best 6 1-lb bottles of run or extracted honey —1st, Mr. Wootton; 2nd, Mr. Southwell; equal 3rd Mr. Mortimer, Oxton
Best specimen of bees with queen in unicomb hive—1st, Mr. Mortimer, Oxton; 3rd, Mr. Brooks

At each show lectures with manipulations were given in the bee-tent by Mr. TB Blow, who also acted as judge in the honey classes.

Some of the Notts. County Councillors put in an appearance at these shows and evidently appreciated the way in which the subsidy received by NBKA from the County Council was being applied.

Rev Henry Lewis Williams became vicar of St Mary's, Bleasby in 1888. He was born in 1843 in Chorlton and died in Southwell in 1911.

BBJ 24 Aug 1893
The annual exhibition of NBKA was held in connexion with the fete of the Porchester Floral and Horticultural Society, Nottingham, at Mapperley, on Bank Holiday, August 7th, and following day.

The show was in every way a great success, its supporters staging, in six honey classes, close upon half a ton of honey, which was excellent in its quality and get-up. Of the five honey trophies staged, three contravened rules of the schedule, and were consequently disqualified. Two of the disqualified exhibits would have taken first and second prizes, and so it was decided to withhold the first and to award the second and third prizes to the two remaining exhibits.

The arrangements and staging of the show reflect great credit upon the Executive of the Association. Mr. Jno. Palmer, of Ludlow, acted as judge and during the Monday afternoon held an examination of candidates for third-class experts' certificates in the

bee-tent, where short lectures on beekeeping were delivered at intervals on both days.
Prize List.
Collection of hives and appliances—1st, AW Pett, Nottingham
Best frame hive for cottagers' use—1st, C. Redshaw, Wigston; 2nd, WP Meadows, Syston
Bees in observatory hive—2nd, HJ Raven, Bridgford; 3rd, Th. Maskery, Kirkby; 4th, AJ Mortimer, Oxton
Honey trophy—1st, not awarded; 2nd, AG Pugh, Beeston; 3rd, HJ Raven
Twelve 1-lb. jars extracted honey—1st, G. Wood, Oxton; 2nd, T. Riley, Radford; 3rd, Ch. Wootton, Draycott; 4th, J. Setchfield, Hyson Green ; 5th, E. White, Bleasby; vhc, AJ Mortimer
Twelve 1-lb. sections—1st, AJ Mortimer; 2nd, AW Pett; 3rd, W. Measures, Upton
Six 1-lb. jars granulated honey—1st, H. Merryweather, Southwell; 2nd, S. White, Bleasby; 3rd, HJ Raven
Best frame of honey—1st, TS Elliott, Southwell; 2nd, W. Measures; 3rd, G. Wilson, Nottingham
Six 1-lb. jars extracted honey (non-prizewinners at previous shows only)—1st, J. Setchfield; 2nd, Th. Maskery; 3rd, T. Riley
Beeswax—1st, HJ Raven; 2nd, AJ Mortimer

BBJ 31 Aug 1893
Country Doctor (Newark). The queen sent is evidently an aged one, probably about three years old but this is in a measure guesswork, as we have but the ragged wings and general appearance to guide us in forming an opinion.

BBJ 7 Sep 1893
Shropshire BKA Shrewsbury Show: During the afternoon on the first day, a meeting of the Northern District Committee of the BBKA was held, under the presidency of the Rev. JF Buckler, the main subject of discussion being the question of providing judges and examiners at county shows. The subject was ably opened by Mr. Pugh, of NBKA, directing attention to the need that existed of appointing a larger number of capable judges in many of the districts. This was admitted by subsequent speakers, and the view was adopted generally that the want could be best supplied by the County Associations putting forward suitable persons to receive guidance and experience by being linked with the present recognised judges. This recommendation it was decided to send up to the Committee of the central Association.

BBJ 28 Sep 1893
The seventh and last Exhibition arranged by NBKA for the season was held at Moor Green on September 5th in superb weather and under favourable auspices, which resulted in a very large attendance. The exhibits were both numerous and decidedly

good, reflecting credit on the Society and its excellent Secretary, Mr. Pugh, who was assisted by the Steward, Mr. A Warner, Mrs Warner, and other friends. Addresses given in the bee-tent by Mr. CN White were listened to by large numbers with evident interest. Mr. White also acted as judge of the honey department.

List of Prizes.
Twelve 1-lb. sections—1st prize, P. Scattergood; 2nd, H. Wiggett; 3rd, M. Lindley
Twelve 1-lb. jars extracted honey — 1st, C. Wootton; 2nd, T. Letchfield; 3rd, JF Simpson
Best exhibit of bees—1st, A. Warner; 2nd, AJ Mortimer; 3rd, H. Wiggett
Best bell-glass super—1st, A. Warner
Best frame of honey in comb — 1st, JF Simpson; 2nd, G. Reeve; 3rd, M. Lindley

BBJ 28 Sep 1893
Width of Shallow Frames in Surplus Chambers. Will you kindly allow me to call the attention of readers of the *Journal*, and especially the attention of our leading bee-men, to the above subject, in the hope that it may lead to a discussion on the matter and, if possible guide our leading bee-appliance makers, and, if needs be, the Committee of the BBKA, in suggesting to the beekeeping fraternity what is the best width to work shallow bars.

In the report on the honey classes at the Royal Show at Chester, July 6th, you said "Nothing in honey production can, to our mind, surpass a well-finished shallow frame of good honey; it far exceeds a single section in beauty." I quite endorse all this, but I would like to add that if they are worked at a distance of two inches from centre to centre, instead of the usual one and half inches, they do surpass sections and, by a very long way, those frames that are worked out at one and half inches. But I suppose here is the crux of the whole matter and about which there seems some diversity of opinion – as to whether the new metal end should be two inches, or more or less. Personally, I favour the two-inch end and a five-inch deep shallow bar; then I can work either WBC. hanging-section crate, or shallow bar, or a combination of both if wished and I have seen in this county (Notts.) a number of frames taken from hives this season and worked with two-inch ends, that were a perfect picture, while the gain to the bees in capping, and to the bee-keeper in time and honey, as pointed out by friend Woodley (p. 304, in Journal of August 3rd) is very great indeed.

In this county one of the members of the NBKA has hit upon a novel plan. He makes the top bar something after the plan adopted in the WBC hanging-section crate and exactly the same width, viz., two inches, the sides or ends of the frames are one inch and seven-sixteenths wide, and the top bar and sides are grooved to receive the foundation; the bottom strip is one inch wide and three-sixteenths thick and the bees have plenty of room to get to their work.

Another plan adopted by a second bee-keeper here for several years past is to place strips of wood one-half inch square and seventeen inches long between each WBC end, and thus make the ordinary bars have two-inch ends; and he tells me that he has had splendid results from this method. The bees have built the combs out the extra width, and, while having less frames in number, he has had a larger quantity of honey.

I could say more, but I think perhaps I have trespassed enough on the editors' space and patience, so leave what I may have to say further to another issue.

Hemlock Stone, September 18th, 1893.

[The maker of the WBC end is, we believe, now engaged on the subject of a wider end for shallow frames which will meet the requirements of the case, and it is certain that before next season such ends will be put on the market.

Personally, we have always taken means to have heavier combs in surplus chambers than the ordinary "end" made provision for, by inserting slips of wood between each frame. Our own idea would be to make the new end of such a width that eight frames would occupy the ordinary shallow-frame box now holding ten frames. We hope to bring the matter up at the next meeting of the Committee of the BBKA.—Eds.]

BBJ 28 Sep 1893

Sending Honey by Post. I am sending by this post two samples of honey on which I should be glad to have your opinion. Sample No. 2 was gathered in August; is it from the blackberry? Why is sample No. 1 so dark? This has been a very good year here; I have taken 274 pounds extracted, and twenty-nine sections from four stocks, the best yielding 101 pounds extracted. Quality, mostly very good; some of the early takes were, however, rather spoiled by honeydew.

With regard to the "Useful Hints" about packing honey for post, I don't think that the postal authorities are altogether blameless in the matter, so far as my experience goes. I packed my best section, sealed all round, with glass each side, in a close-fitting postal section-case, carefully labelled as to its fragile nature and dispatched the same to a show for competition; it was broken in transit, and unfit to stage. On another occasion I packed a bottle of honey in thick corrugated paper and that also was broken in the post. I trust, however, that the bottles I am sending to-day will be mercifully dealt with.
Percy Sharp, Newark-on-Trent.

[No. 2 is very good clover honey. No. 1 also good but a little spoiled by honey-dew. Bottles arrived safe, but were not very well packed. Two coils of corrugated paper would have been better than one. Eds.]

Dr Percy Sharp(e) was born in Bradford in 1866. He appears many times in this narrative in connection both with local and national beekeeping matters. He lived in Brant Broughton in

BEEKEEPING IN VICTORIAN NOTTINGHAMSHIRE

Lincolnshire towards the end of the century. In 1896 he was appointed (alongside JC Cribb) as expert to the Lincolnshire BKA. In 1905 he is recorded as taking part in motorcar races in a 12hp Richardson vehicle.

BBJ 28 Sep 1893
Overdosing with Salicylic Acid. I have recently bought a stock of bees to start beekeeping and in making autumn food I have put half an ounce of salicylic acid, instead of half an ounce of salicylic acid solution to five pounds of sugar and I have already given it to the bees which have consumed about four parts of it. I did not know what a great mistake I had made till to-day when a friend called and told me to write to you as quick as possible. Kindly give me your advice as to what I should do.
JWR, Burton Joyce, Notts.

Reply. There is no help, of course, for what has been done and the overdosing, though bad enough, is not quite as bad as it seems because only a certain portion of acid can be contained in solution and the rest would be wasted. Use no acid in the rest of the food.

BBJ 19 Oct 1893
GH Young (Notts.). Location for Bees. Much will depend on the quantity of white clover growing within one to two miles of the place the bees are located. Heather is always helpful to the honey-crop, but we do not think it produces very much about Bournemouth. It always comes in after clover has done yielding.

BBJ 4 Jan 1894
At the meeting of the BBKA Committee, held at 105, Jermyn-street, on December 20, the Secretary reported that he had written to the several affiliated Associations urging them to solicit the support of the several Agricultural and Horticultural Societies within their respective districts and that Nottinghamshire had acknowledged the receipt of this communication.

The Chairman reported, on behalf of the Educational Sub-Committee, that Dr. TS Elliott, of Southwell, Notts. had passed the second-class examination.

BBJ 1 Feb 1894
G. Mills (Newark). Dead Bees cast out. There is nothing in the bees sent to indicate the cause of death. No alarm, however, need be felt at the occurrence. There would be no actual "harm in giving candy" now but if food is plentiful it would do no good, and consequently is unnecessary.

BBJ 15 Feb 1894
The Notts. County Council have arranged with the NBKA for lectures on bee-keeping to take place as under: -

February 14, Retford, Temperance Hall, 7.30.
February 15, Newark, Old Savings Bank, 7.30.
February 16, Stapleford, Board School, 7.30.
February 17, Nottingham, People's Hall, 6.30.
February 20, Epperstone, National School, 7.0.
February 21, Kirkby, National School, 7.30.
February 27, Granby, National School, 7.0.
February 28, Ruddington, National School, 7.30.
March 1, Clifton, National School, 7.0.

The first four lectures of the series will be delivered by Mr JH Howard; those at Epperstone and Kirkby by Mr. AG Pugh; at Granby and. Ruddington by Mr. JH Raven; and the final one, at Clifton, by Mr. P Scattergood, jun. In addition to the above, lectures have already been given at Bleasby and at Strelley by Mr. G. Hayes and Mr. P Scattergood, jun.

Lincolnshire Echo. 15 Feb 1894
Last night, under the auspices of the Technical Committee of the Notts. County Council, Mr. John H Howard gave a lecture in Retford on elementary beekeeping. Mr. Howard's remarks, which were of an interesting character throughout, were illustrated by lantern views. The Mayor (John H Hewitt) presided, and there was a fairly large attendance.

Minute Book - The NBKA Annual Meeting was held at the Peoples' Hall, Heathcote-street, Nottingham on 17th February 1894. Although it was a most inclement day there was a large number of members present when the President (Viscount St. Vincent) took the chair at 3.10 pm. His Lordship, having opened the meeting, requested that the Secretary read the minutes of the last Annual Meeting which, having been done, Mr. Roe proposed and Mr. Rawson seconded, that the same be confirmed.

The Secretary then read the Annual Report which was of a satisfactory character recording a steady increase in membership which now, for the first time in the societies' existence, exceeds two hundred members. It also congratulated members on the past year having been a fairly good honey season. Viscount St. Vincent proposed and Mr. Rawson seconded that the annual report as now read be adopted and that the same be printed and circulated to members.

The Hon. Secretary read the annual balance sheet comparing the same with the previous one. Mr. Scattergood proposed and Mr. Marriott seconded that the annual balance sheet has now read be adopted.

BEEKEEPING IN VICTORIAN NOTTINGHAMSHIRE

The election of the committee was then proceeded with and the following gentlemen were elected members:
J Baguley, FHK Fisher, M Lindley, JW Marriott, W Poxon, JT Faulconbridge, J Finn, C Forbes, J Rawson, Snr, G Wood, GH Elliott, G Webster, H Merryweather, H Miller, and J Burton.

Mr. Baguley proposed and Mr. Rawson seconded that Mr. P Scattergood be thanked for past services and be re-elected as Auditor. Mr. Roe proposed and Mr. Riley seconded that Mr. AG Pugh be re-elected hon. secretary and treasurer on the same terms as last year. Mr. Pugh proposed and Mr. Marriott seconded that Viscount St. Vincent be most heartily thanked for past services and be invited to accept the position of President for 1894.

Mr. Roe proposed and Mr. Riley seconded that Messrs. Pugh and Scattergood be representatives to BBKA for 1894 and that an allowance of 10/- each be paid towards their expenses for each meeting they attend in that capacity.

A discussion took place as to holding the Annual Show in 1894 and it was decided to leave the matter as well as prize schedules to the committee. It was mentioned by the Secretary that the BBKA had decided to withdraw their medals for 1894 and the President kindly offered to pay for them if the BBKA would allow him. The Secretary to write and ascertain if they were willing to allow this.

Tea was then partaken of by those present. After tea the President presented Mr. Wood of Oxton with a silver medal, Mr. Mortimer of Oxton with a bronze medal and Mr. Merryweather of Southwell with a certificate - all of which were won at the last Annual Show in Porchester. He also very kindly distributed prizes for best honey yield as follows:
Best yield of honey from apiaries of 5 hives upwards
    1st 15/- Mr. Measures, Upton
    2nd 5/- Mr. Raven, Bridgford
Best yield of honey from apiaries of less than 5 hives
    1st 15/- Mr. Maskery, Kirkby
    2nd 5/- Mr. Fleet, Bramcote
Best yield from a single hive
    1st 10/- Mr. Measures, Upton
Best single bottle of granulated honey shewn at Annual Meeting
    1st 10/- Mr. Elliot, Southwell
Best single section of honey shewn at Annual Meeting
    1st 10/- Mr. Scattergood, Stapleford
It was proposed, seconded and unanimously adopted that the Committee be requested to design and register a label for members to use when selling their honey,

rules under which this is done to be settled and labels forwarded in time for current season harvest.

Mr. JK Howard then gave his lecture on the "Two Queen System" and "Marketing Honey" which was followed by discussion and the annual prize drawing when prizes won were as follows:

1st	Wells hive presented by the President	Mr. Letchfield, Radford
2nd	XL Hive presented by the President	Mr. Robinson, Radford
3rd	Hendon Extractor	Mr. Finn, Basford
4th	Hive	Mr. Wood, Oxton
5th	Feeder presented by Mr. Marriott	Mr. Ball, Hucknall
6th	Feeder presented by Mr. Scattergood	Mr. Baguley, Chilwell
7th	100 Sections presented by Mr. Pugh	Mr. Roe, West Bridgford
8th	Volume of "Record"	Mr. Higgett, Hucknall
9th	Volume of "Record"	Mr. Housley, Beeston
10th	Buxton feeder presented by Mr. Puttergill	Mr. Davies, Bagthorpe

Meeting terminated at 8.45pm.

George E Puttergill was a woodwork teacher in Beeston. He was born in 1865 in Caunton.

NG 24 Feb 1894
The annual general meeting of NBKA was held on Saturday afternoon in the People's Hall, Heathcote-street, Nottingham. The president of the association, Viscount St. Vincent, occupied the chair. The Secretary read the annual report. It was gratifying to record that there was a still further increased membership. During the year four members had left the county, eight who lived out of the county had ceased to subscribe and 22 had resigned, but against this total loss of 34 had to be placed 42 new subscribers who had been enrolled. The membership at the close of 1893 was 203, against 195 at the same time in 1892. (Applause)

The committee regretted that they had to show a small sum owing to the treasurer. After very serious consideration they felt convinced that this was brought about by many members paying cottagers' subscriptions who really were not cottagers in any sense of the word. If only half of those who were now paying that subscription would increase it to the proper amount in future the adverse balance would disappear and there would be no need to curtail the present privileges of membership - a course they would have to adopt if this much-needed reform did not take place. He {Mr. Pugh) was glad to say that several members had already increased their subscriptions from 2s 6d to 5s. (Applause.)

The honey returns for which their noble President so kindly offered liberal prizes had not been as numerous as was expected. This was to be regretted because of the scant encouragement that had been given to his Lordship to continue his liberality in this direction and also because those returns would have proved so valuable in computing the actual amount of honey gathered in the county. It was, however, very pleasing to state that those who had sent in returns showed a considerable increase on the previous year's yield, the average per hive all round having advanced from 27 lb to 40 lb, or 13 lb per hive. It was fair to assume that the members, who owned at least a thousand stocks, would harvest 10,000 lbs more honey in 1893 than in 1892 and as these would be worth over £400, they might congratulate themselves upon fact that 1893 had proved one of the honey years they had been looking for a long time. (Applause.)

Some discussion ensued as to where the next annual show should be held and eventually it was decided to refer the matter to the committee. The Secretary stated that an intimation had been received from the BBKA that the silver and bronze medals would be withheld through 1894 on account of financial considerations. The Chairman said that if the BBKA would let them have the medals on paying for them he would be glad to pay for them. (Applause.) The report was then unanimously adopted on the motion of the Chairman, seconded by Mr. Rawson and the secretary was asked to communicate with the BBKA on the matter referred to by the Chairman.

Messrs Pugh and Scattergood were appointed to represent the association at the quarterly meetings of the BBKA. The following officers were elected: — President, Viscount St. Vincent; vice-presidents, the Duke of Portland, Lord Newark, MP, Mr. JE Ellis, MP, Ald. SH Sands, the Rev. Watkin Homfray, Mrs. AH Chambers, and Mrs. J. Hind. Committee: Messrs. J Baguley, M. Lindley, SW Marriott. W. Poxon, JT Faulconbridge, J. Finn, H. Merryweather. C. Forbes, J. Rawson, sen. T Miller, WH Webster, H. Newton, G. Wood and the district hon. sec, hon. treasurer, and secretary, Mr. AG Pugh.

Afterwards the members sat down to a meat tea, and this was followed by a lecture by Mr JH Howard, on "The Two Queen System" and "Marketing Honey" Mr. P Allsebrook, of Wollaton, presiding. The medals and certificates won by the members during the year were distributed, and the annual prize-drawing took place. Prizes had been offered by the president for the best honey returns

BBJ 15 Mar 1894
The first meeting of the BBKA Committee elected for the ensuing year was held at 105, Jermyn-street on the 7th inst. Mr. H Jonas was voted to the chair.

Letters were read from the Secretary of the Notts. Association in reference to the decision of the Committee (and subsequently confirmed by the general meeting)

requiring the affiliated Associations to make an extra payment for medals and certificate if offered for competition during 1894 and the adoption of the recommendation made by the Northern Associations in regard to an increased payment in entrance fees by candidates when competing for first, second, and third class certificates.

The Secretary was requested to inform the Notts. Association that the Committee had given much consideration to these matters and that in their recommendations they had endeavoured to minimise as much as possible the difficulties the Notts. Association supposed would arise in the arrangements which had been made. Mr. P Scattergood, jun., representative of the Notts. Association, was approved as an *ex-officio* member of the Committee.

BBJ 15 March 1894
Mouldy Pollen in Store Combs. On looking over my stock of empty combs in the shallow extracting frames to get them ready for the coming season, I have found that there is a considerable quantity of pollen in some of them and some of the pollen has gone mouldy. I thought that it would save foundation if I were to remove the cells and pollen as completely as possible down to the foundation and then let the bees repair these combs in the summer. I found, however, that it was impossible to remove all the pollen. Would you kindly advise me, therefore, in the Bee Journal whether a little pollen lying about would be removed by the bees, or would it be left and so spoil the colour and flavour of honey stored in those combs? Had I better melt down those combs for wax and give the bees fresh sheets of foundation? Thanking you in anticipation of a reply in an early issue of the Journal.
       "Country Doctor," Newark, March 7. (Is this Percy Sharp?)
Reply. In dealing with such combs, other than by melting down, we have never had satisfactory results. Far better to have new combs built as proposed.

Minute Book – The quarterly meeting of NBKA was held on 7th April in the People's Hall, Heathcote-street. Viscount St. Vincent in the chair. Present were Viscount St. Vincent, Messrs Burton, Elliott, Faulconbridge, Finn, Fisher, Marriott, Merryweather, Miller, Poxon, Rawson, Webster, Wood, Scattergood, Gosling, Warner, Newton, Riley, Hallam and the Hon, secretary, AG Pugh.

Mr Riley proposed and Mr Finn seconded that the minutes of the last quarterly committee meeting as now read be confirmed.
Mr Marriott proposed and Mr Fisher seconded that the quarterly balance sheet as now read be adopted.

The Secretary submitted four designs for the proposed honey label and a long discussion took place as to the advantages likely to be derived from their use, the President being of the opinion that their use would have a tendency to bring prices down to a dead level, unless great care was advertised that the annual meeting had sanctioned them. It was, however, to at once bring them into use and Mr Derry's design, with the addition of a line for name of seller and an estimate for 20,000 was accepted. The question of appointing association agents was left to the Hon. Sec. to arrange or bring before a future meeting.

A discussion then took place upon correspondence read by the Hon. Sec. re. 'place for annual show in 1894'. Mr Elliott proposed and Mr Merryweather seconded that the Southwell Horticultural Society's invitation offer of £15, tenting, staging etc., at their show on July 26th, 1894 be accepted.

Dr Elliot proposed and Mr Newton seconded that the Secretary obtain a quotation for a pendant to be used as special prizes at our annual shows.

The prize schedules for the annual show was considered, Lord St. Vincent very kindly offering to give similar prizes to what he did year. The Hon. Secretary was instructed what prizes to offer in each class and to insert a clause in the schedule to the effect that two members might combine to show in the Trophy class providing that they did not run more than six stocks of bees at the time of entry. It was also decided to allow winners of first prizes to take pendants in lieu of cash where the prize was of sufficient value and they preferred it. A clause re. 'protests' was also agreed.

The meeting terminated at 5.45pm.

NG 14 Apr 1894
Continuation Schools Conference in Nottinghamshire. With a view of ascertaining the opinions of school managers and school masters in the county on the subject of the development of the continuation school movement and the best means of securing such development, the Technical Instruction Committee of the Notts. County Council arranged for a conference, held at the Shire Hall, Nottingham, on Saturday afternoon. There was a large attendance, Lord Belper, president of the County Council, in the chair.

Beekeeping lectures and demonstrations have been given under the auspices of NBKA, to which a grant of £40 had been made for this purpose.

BBJ 19 Apr 1894
Meeting of the BBKA committee, held at 105, Jermyn-street, on 12th inst., Mr. H Jonas (vice-chairman) in the chair. Communication were received from Mr. P. Scattergood, jun., regretting inability to be present

NEP 24 Apr 1894

It is proposed by the Notts. County Council to make a further grant of £20 to NBKA for a series of lectures and demonstrations in beekeeping.

BBJ 3 May 1894
Suspected Foul Brood. I have a stock of bees affected with foul brood. The stock is fairly strong and very active now, and the combs are fairly well filled with brood. The foul brood can be detected by the eye and also by the bad smell. If the disease is not dealt with I am afraid it will spread during the summer and get worse. Please give your advice as to what I had better do. JES per HS, Arnold.
Reply. We much prefer being certain as to the presence of foul brood before advising. Kindly cut out a piece of the comb containing dead brood and forward it in a tin box—or some such package—when we will be better able to deal with the case. So many mistakes are made in judging whether brood is diseased or not, we cannot do less than ask that this should be done.

BBJ 6 May 1894
The annual general meeting of the NBKA was held on February 17, in the People's Hall, Heathcote-street, Nottingham. The President, Viscount St. Vincent, occupied the chair, and there was a good attendance.

The Secretary read the annual report, which it is gratifying to note recorded an increase in the membership, the numbers being 203 against 195 in the same period of 1892. It was matter for regret that the accounts showed a small balance due to the Treasurer, owing, it was thought, to the fact of many members paying cottagers' subscriptions who really were not cottagers in any sense of the word.

The committee hoped the members referred to would recognise the necessity for altering this state of things, and avoid the need for a curtailment of their present privileges, which must otherwise result. He (Mr. Pugh) was glad to say that several members had already increased their subscriptions from 2s 6d to 5s. After detailing several matters connected with the past season's honey results and the various shows held by the Association, the report concludes with some particulars of the bee lectures given under County Council auspices which have already appeared in our pages.

The Secretary stated that an intimation had been received from the BBKA that the silver and bronze medals would be withheld through 1894 on account of financial considerations. The Chairman said that if the BBKA would let them have the medals on paying for them he would be glad to pay for them. The report was then unanimously adopted, and the Secretary was asked to communicate with the BBKA on the matter referred to by the Chairman.

The Secretary then presented the annual financial statement. They had 132 subscribers of 2s 6d and only 45 of 5s. The amount received in subscriptions was £38 13s; donations to prize fund, £32 2s 6d; these, with other items, including £2 17s 2d balance due to the Treasurer, made up total receipts of £80 11s 11d; £33 7s was expended in prizes, and on the year's working there was a loss of £1 8s. The Chairman said they had expended practically the whole of their money among the members of the Association. They had given £33 in prizes against £23 last year. Of course it was an unfortunate thing to have a little debt, but, as a matter of fact, it was quite as good for themselves as having a big balance in hand.

The balance-sheet was then passed, with a vote of thanks to the Auditor (Mr. Scattergood), who was unanimously re-elected, Messrs. Pugh and Scattergood were appointed to represent the Association at the quarterly meetings of the BBKA. The following officers were elected:—President, Viscount St. Vincent; Vice-Presidents, the Duke of Portland, Lord Newark, MP, Mr. JE Ellis, MP, Ald. SE Sands, the Rev Watkin Homfray, Mrs. AH Chambers, and Mrs. J. Hind. Committee— Messrs. J. Baguley, M. Lindley, SW Marriott, W.Poxen, JT Faulconbridge, J. Finn, H. Merryweather, C. Forbes, J. Rawson, sen., T. Miller, WH Webster, G. Wood, and the District Hon. Sec, Hon. Treasurer, and Secretary, Mr. AG Pugh.

The business part of the proceedings terminated with a vote of thanks to the Chairman. Afterwards the members sat down to a meat tea and this was followed by a lecture by Mr. JH Howard, on "The Two-Queen System" and "Marketing Honey," Mr. P. Allsebrook, of Wollaton, presiding. The medals and certificates won by the members during the year were distributed, and the annual prize drawing took place.

BBJ 31 May 1894
Notes from Notts. I was pleased to see your editorial remarks respecting foundation, as I am sure that very many bee-keepers do not discern the difference in the way the cells should run. The thin-wired foundation imported from America some years ago was cut in the wrong direction and I was much surprised to see a lot of foundation supplied by one of our best appliance dealers this year also cut so that the parallel sides are horizontal instead of vertical.

I have often expected to see some of your correspondents ask why entries for the "Royal" shows have to be made such an absurdly long time before the date on which the show takes place. Even post-entries at double fees close on May 12, nearly a month sooner than ordinary entries might reasonably be accepted. I feel certain this must, especially in good years, tell against the show being more popular. We in Notts. always try to keep entries open till as near shows as possible, and I think it might be done with all-round advantage in the case to which I have drawn attention.

Following the good example of Berks BKA we have adopted Association honey labels, a sample of which I submit for your inspection. AG Pugh, Hon. Sec. NBKA [Referring to our correspondent's complaint of the early date on which entries for the "Royal" shows close, he loses sight of the enormous extent of the preparations required for what is justly known as the largest annual Agricultural Show held in the world. Space has to be apportioned and allotted a long time beforehand, the official surveyor requiring a plan of each department and the space required many weeks before the show takes place, in order to facilitate the arranging of the ground, as well as the preparation of catalogues, &c. All this is unavoidable in a show of such magnitude as the "Royal." Beekeepers, however, have the privilege of claiming the return of their entrance-fees if, owing to bad weather, their exhibits cannot be completed in time and this arrangement should go a long way towards removing the hardship of closing entries so long beforehand. Eds.]

BBJ 7 Jun 1894
May 26. NBKA Annual County Show at Southwell. Entries close July 19. Secretary, AG Pugh, Mona-street, Beeston. Other shows connected with the NBKA will be held as follows:
Lowdham, July 19; Hucknall Torkard, July 24; Beeston, August 6; and Moorgreen, September 4.

Leicester Chronicle 11 Jun 1893
Under conditions which warranted the anticipation of considerable success, the fifteenth annual show of the Nottinghamshire Agricultural Society was opened yesterday in Colwick Park. Never has the county organisation found itself in a position which gives cause for greater encouragement. It was altogether a delightful day, bright, warm, and genial, and in the park it was delightful for people to stroll about. Demonstrations of beekeeping have been arranged for daily.

BBJ 21 Jun 1894
Requeening and Buying Queens. As one often sees both advice and warning in your journals on the above subject, it occurs to me that my bitter experience may induce some people to "let well alone," as I intend to in the future ; at any rate as regards buying queens.

Last autumn I desired to requeen two hives; I therefore bought a queen, for which I gave 8s 6d and having removed the old queen from No. 1, I introduced the young one by Simmins' direct method, and was perfectly successful in the operation. Thinking 8s 6d rather a high price to pay. I decided to buy a cheaper one for No. 2, and accordingly bought a queen for 2s 6d from another well-known dealer. On receiving this, even I, a comparative novice, could see that she was old and no better than the one I had removed but to make certain I showed her to an expert at a show. He asked me where I got her and on hearing the name he said "she was sure to be all right from such a source," but did not examine her.

I, however, decided not to adopt his advice and bought another from the first dealer, giving 7s 6d for this one, and introduced her safely. A few days after that an expert visited me and I showed him the queen I considered old, without saying where I had obtained her. He examined her carefully and said that she was "not less than three years old." In order to be doubly sure, I sent that queen to you, sir (dead), and you kindly replied in the Journal, "Queen sent is evidently an old one."

I had requeened No. 2 on August 4. On August 22 I again found them queenless, possibly from being handled too soon after requeening. I therefore wrote to a third dealer for another young queen, for which I paid 3s 6d, the receipt describing it as "one 1893 fertile queen." She was safely introduced, and the hive fed and packed up for winter. On April 10 last I made my spring examination. No. 1 was all right but in No. 2 I found no worker-brood whatever but a large quantity of drones and drone-brood. I concluded that the hive was queenless (as I could not find her) and had a fertile worker. I therefore gave them a comb of eggs and brood from my strongest hive (which latter "balled" its queen in consequence) but they raised no queen cells.

Then we had a long spell of bad weather and I was unable to do anything for them and the drones increased in numbers while the workers diminished. At last, on May 25, I examined them thoroughly; there were hundreds of drones, a score or so of workers, and a queen.

Of course, the stock was useless, so I removed the queen and sent her to the chief expert of my association (with whom I had already had some correspondence on the subject) and he replied that the queen was "an old drone-breeder," and being rather small, I had failed to find her. The hive that had "balled" their queen had raised another, which is now laying, I am glad to say.

I do not think that it is quite honest of dealers to advertise young queens at 2s 6d and 3s 6d if they intend supplying old ones to people whom they think are novices.

PS, Newark. June 12. (Percy Sharp?)

Minute Book – Quarterly Committee Meeting held on July 7th 1894 at the Peoples' Hall, Nottingham.
Present – Messrs GH Elliot (Chairman), Baguley, Faulconbridge, Forbes, Miller, Poxon, Rawson, Webster, Scattergood, Lane, Warner, Newton, Hallam, Turner and AG Pugh (Hon. Sec).

Mr Scattergood proposed and Mr Webster seconded that the minutes of the last committee meeting as now read be hereby confirmed.

Mr Scattergood proposed and Mr Baguley seconded that the balance sheet as now read be adopted.

The Secretary read various correspondence and some discussion took place with regard to one from Mr J Garrett respecting Third Class examinations. The feeling of the meeting was that although it is desirable that examination of these candidates should not interfere with the judge's duties, yet the shows being most convenient days for candidates to attend, the second day of two-day shows would be a most suitable time. It was also suggested that if each county cannot find sufficient candidates that examinations should be held alternate years at the Notts, Derby and Leicester Shows unless sufficient candidates ask for a special centre to be selected to meet their convenience. The Secretary promised that their ideas should be brought to the notice of the BBKA committee.

The Secretary complained of the great difficulty he had experienced in getting anyone to act as steward at the Hucknall Show. It was suggested that, unless greater willingness is shown in that direction another year, it will be advisable to curtail the work done.

The Secretary also informed the meeting that Mr Pett had suggested showing at the Annual Show and asked for instructions. Mr Scattergood moved and Mr Newton seconded that if Mr Pett pays 5/- and recalls any objectionable remarks, he be allowed to show as usual.

The Secretary read various correspondence and some discussion with regard to one from Mr J Garrett respecting Third-Class Experts.

Mr Webster moved and Mr Newton seconded that Mr Webster of Binfield, Berks. be judge at the Annual Show at Southwell on July 26th 1894 and also at Hucknall. Mr Scattergood moved and Mr Forbes seconded that Mr Turner of Radcliffe be Assistant Judge at Southwell Show at an inclusive fee of 7/6d. Mr Scattergood moved and Mr Miller seconded that Mr Turner be asked to act as judge at the Beeston Show at the usual fee of 5/-. Mr Turner moved and Mr Miller seconded that Mr P Scattergood be asked to act as judge at the Moorgreen Show at a fee of 5/-.

The Secretary submitted a new honey label which was approved of. It was decided to leave the question of appointing agents until a later date.

Foul brood remedies were mentioned. It was decided to leave the purchasing of some for members use to be dealt with at a further meeting.

BEEKEEPING IN VICTORIAN NOTTINGHAMSHIRE

BBJ 12 July 1894

"Re -Queening And Buying Queens." In reply to Mr. HW Brice's letter (p. 264) I wish to thank him for his friendly criticism of mine on p. 245 on the above subject. I regret having made such a "hash of my bees" as Mr. Brice says I have, but I distinctly said in my letter that I should "let well alone . . . . as regards buying queens" so that I hope not to be an instance of the old adage quoted by Mr. Brice. Thanks to Mr. Brice I now know that 7s 6d is too much to give for home-bred queens. Though my case was not a parallel to the one quoted, as it was not the high-priced queens I found any fault with, only the cheap ones.

To answer Mr. Brice's questions: I did not destroy the old queen of No. 1 until I had made certain that the introduction of the other had been safely effected. The first queen bought for No. 2 died of starvation in the box she was sent in, as I had decided not to use her and consequently did not attend to her wants.

As regards the death of the second queen bought for No. 2 in eighteen days (or less), I suggested in my letter that I thought that hive was examined too soon after her introduction, as I received a visit from an expert on August 11, and he examined all my hives.

Mr. Brice takes exception to my not examining the hive thoroughly before May 25 and preaches me a little sermon on the text of thoroughness in all we do. Again I thank him; but I had a reason, which was that in thoroughly examining them in early spring I was in fear of chilling the brood (if there was any) and also of having the queen balled if she were in the hive.

I sincerely hope that I shall not fulfil his gloomy prognostications, and "go from bad to worse," but that I may with experience and thoroughness gain that skill and knowledge of which it is evident Mr. Brice has so large a share.—PS, Newark, July 2.

BBJ 12 July 1894

Swarming Mishaps. The stock of bees first referred to in my letter of the 12th ult. on re-queening as doing well swarmed on June 8; but being a dull day the swarm returned to its own hive. The following days were cold and wet until the 14th, when the swarm again issued, and, after hiving in a skep, was placed on a board. An hour or two later my groom noticed a considerable number of bees flying round the skep.

Later I inspected them, and found large numbers of dead and dying bees on the board and under the skep. We therefore at once prepared a frame-hive, having in it fully worked-out but empty combs, placing it in the position occupied by the hive which had swarmed, and which we moved a little distance away, in order to strengthen the swarm with the flying bees. We then hived the swarm, and, putting a feeder on, left them. In the evening we weighed the dead and they amounted to $1/2$ lb. This morning I saw a large number of bees around the old hive, and, on inspecting the hive in which I had put the swarm, found it deserted, except for a very few bees, which were cleared out, and the hive closed. We then inspected the old hive and found it crammed with bees and on the only frame I took out were two queen cells.

1. In front of the old hive we found the enclosed dead bee. It looks too small for a queen, but not the shape of a worker. Which is it?
2. Could you tell me the cause of death of so many of the swarm? PS, Newark.
Reply. 1. Bee received is a young queen.
2. We can only suppose that a partial attempt at swarming by the bees of another hive has occurred simultaneously with the issue of swarm referred to, and, both lots of bees having united in one cluster, a fight resulted. If in the melee the queen accompanying the swarm got killed, it would, of course, account for the bees returning to the parent hive. It seems clear, however, that preparations for swarming had been made prior to the 8th, when the top swarm actually came off, and that the old queen met with some mishap on that day. Otherwise there would not have been a young queen hatched out six days later, viz., on the 14th.

BBJ 19 Jul 1894
The monthly meeting of the BBKA committee was held at 105, Jermyn-street, on 12th inst. TW Cowan (in the chair). Resolved that the following Midland and Northern affiliated associations including NBKA be invited to send a representative to attend a meeting to be held at Shrewsbury on August 22, so that the subject of centres and arrangements for third-class examinations be taken into consideration.

BBJ 26 Jul 1894
"In a Fix" (Notts.) has sent a sample of comb all of which are affected with foul brood. In every case we advise the prompt destruction of all combs with dead brood in them, and if our correspondent will refer to our issue of June 21 last, p. 241, they will find details of the plan we advise for dealing with their respective cases. It is no use trying to save or utilise weak lots of bees from foul broody hives, but if they are fairly numerous — covering, say, six or seven frames — they may be saved.

NG 27 Jul 1894
HAG. (Nottingham) asks what we would advise to do with a small cast and if we think it is worth keeping separate, or would we advise him to unite it with another stock that has an old queen. If you do not wish to increase the number of your stocks it would perhaps be best to take out the old queen from the stock at once and unite the two. If, however, you wish to increase your number and encourage the queen to lay by gentle stimulative feeding for a few weeks, you may build up a fairly good stock to winter, but this is rather risky, because you say the queen in the stock is

an old one. If this is so she might die during the winter and thus leave your stock queenless in the spring time. Under these conditions we would advise you to unite the two after having first deposed of the old queen. In doing this, do not fail to spray both stocks with a thin syrup made by dissolving a few lumps of sugar in about a gill of warm water, adding two or three drops of essence of peppermint. This will cause both stocks to have the same scent, and the bees will unite without fighting. Use the syrup cold, or not more than new milk warm.

EC. (Eastwood) writes: How can I put a new young queen into a hive which at the present time has an old one? There are several methods of queen introduction. First, whichever method you adopt, certain precautions and great care are necessary, otherwise your young queen may be killed by the bees of the stock to which she is introduced. Each hive has a distinctive odour, and in order to introduce an alien queen successfully you must adopt a means of giving the queen an opportunity to form an acquaintance with the bees and acquire the scent peculiar to the hive before being liberated. The old queen having been removed, place your new queen, as you call her, in a queen cage, of which there are various kinds. The Raynor pattern, which is made of perforated zinc, is pushed down between the combs. After the queen has been imprisoned for 24 hours she is liberated by pressing down a wire which opens the bottom part of the cage without disturbing the colony. The Howard pattern, as made by JH Howard of Holme, Peterborough, is really an adaptation of what is known as the "Benton" travelling box and is made of wood. It has a perforated metal lid or slide, and three compartments, two for the queen and a few attendants and the third to hold a supply of what is called "Good's" candy, which is made by mixing honey and fine sugar to a constituency of stiff putty. We have used the cage many times during the last three or four years with uniform success.

If you decide to use this cage place it on the top of the frames over the opening between the centre of the frames and snuggly covered up. Leave it alone till the following evening. Then slide the box only (not the metal covering) as far as it will go. This operation will open the feedhole and through it a number of bees will eat their way through the candy to the queen. Thus in a short time a union with the stranger is affected and she moves out to her new duties. Do not interfere with the hive for two or three days.

Another method, called the direct introduction method of Mr. Simmins, is as follows: Having made your stock queenless, close up the hive, and do not open again till evening when all the bees are in the hive. Just before dusk the queen to be introduced must be taken alone and placed in a matchbox or other receptacle and carried in the trouser pocket or under the waistcoat, where she will be free from chill and kept there for about half an hour. Then go to the queenless hive and give it a slight puff of smoke between the combs and let the queen run into the hive. Return the quilts and coverings and replace the roof and make no examination for 24 hours or more.

NG 28 Jul 1894
Hucknall and District Flower Show. The 21st exhibition of this society was held on Tuesday, in dull weather, which slightly affected the attendance, otherwise the show was excellent in every way. There was a moderate number of entries in the bees and honey department open to members of NBKA, the prize-takers being Mrs. Hind and Messrs. Elliott, H. Cartledge, and W. Lees. Mr. Webster, of Berkshire, was the judge in this section and Mr. WH Cartledge the steward.

BBJ 2 Aug 1894
North Notts. July 30. My hives—seven in number—came out very strong and fresh last spring. No feeding required, but the weather only settled down just in time to prevent me having to do it. Clover honey is our mainstay, but for the first ten days or a fortnight the bees could do nothing; the last week of June and beginning of July, however, made amends. Since then we have had a fortnight's bad weather; very disheartening to see such abundance of clover bloom and the bees not able to work on it.

However, it has again taken a turn for the better; we may get a little more surplus, but if not, my yield will be a fair one, I think. I have not done much extracting as yet, but the quality is everything one could wish. Only one swarm from seven stocks, and that was returned after cutting out queen cells, as I did not want increase. There are very few frame-hives in this neighbourhood —I know of none within several miles. Skeps as a general rule swarmed late, and often only once, and—probably due to the weather—many of them not at all.   Stratton.

Advert Special Strain Virgin Queens for August, from stocks that have given 100 lb surplus this season, 3s 6d each. Webster, Middle Pavement, Nottingham.

BBJ 9 Aug 1894
The annual show of the NBKA was held on July 26, in connection with the Southwell Horticultural Society. The association provided a very full list of prizes both for honey and appliances. In the appliance classes, however, there were no entries. This, no doubt, arose from the fact that one of the clauses in the conditions of exhibition was as follows: — "The first prize will be withheld unless three or more exhibits are staged." A manufacturer would not go to the expense of making an exhibit upon this chance, considering that the second prize was only 20s. There were also several other shows being held at about the same time. The honey classes were well filled with exceptionally fine samples of honey. There were the large numbers of nineteen entries in one class and sixteen in another.

The class for observatory hives with bees was well patronised, there being seven entries. The stewards, who performed their work admirably, were Messrs. GH Elliott (Southwell), J. Holmes, and W. Measures, with the indefatigable Mr. AG Pugh as hon. secretary. The bee-tent (open free, as it always ought to be) was well attended. Lectures were given by a well-known expert from the BBKA who, in conjunction with one of the stewards —Mr. Holmes—acted as judge.

Best Exhibit of Honey in any form, not to exceed 12 lb— 1st, AG Pugh; 2nd, GH Elliott; 3rd, W. Lee

Best Twelve 1 lb. Jars Extracted Honey — 1st, Mrs. White; 2nd, W. Lee; 3rd, Geo. Wood; 4th, T. Maskery; 5th, H. Cartledge

Best Six 1 lb. Sections—1st, H. Wigget; 2nd, G. Marshall; 3rd, WG Elliott

Best Six 1 lb. Jars Granulated Honey — 1st, H. Merryweather, jun.; 2nd, A. Mortimer; 3rd, T. Riley

Best Frame of Honey for Extracting—1st, H. Raven; 2nd, A. Mortimer; 3rd, GH Elliott

Best Six 1 lb. Jars Extracted Honey (previous non-winners only)— 1st, GE Mills; 2nd, J. Herrod; 3rd, G. Bell

Best Three 1 lb. Sections (previous non-winners only)— 1st, WG Elliott; 2nd, Mrs. Harrison

Best Glass Super— 1st, A. Mortimer

Mr. TB Blow's Special Prizes.

Best twelve 1 lb. Sections in Blow's Sections — 1st, A. Mortimer; 2nd, WG Elliott

Best Twelve 1 lb. Jars Extracted Honey in Blow's Jars— 1st, E. Mackender; 2nd, W. Measures

Beeswax— 1st, H. Raven; 2nd, W.Broadbury

Bees with Queen in Observatory Hive — 1st, H. Merryweather, jun.; 2nd, A. Mortimer; 3rd, W. Measures; 4th, G. Wood

NG 11 Aug 1894

Beeston Flower Show. As usual a portion of the large tent was allotted to NBKA who offered prizes in connection with the beekeeping industry, this being one of a series of shows which the Association is holding in conjunction with horticultural societies this season. Mr. AG Pugh Hon. secretary of NBKA, staged for exhibition purposes a well arranged "trophy" of honey. Prizes were offered for 12 1-lb sections of comb honey and 12 1-lb sections of new or extracted honey, specimen of bees (any race) to be exhibited living with their queen in a unicomb observatory hive and for the best exhibit of beeswax. Among the prize winners were H. Cartledge (Hucknall), G. Wood (Oxton), J. Lee (Southwell), Mrs. White, WG Elliott (Southwell), H. Widgett {Hucknall), H. Raven (Bridgford) and AG Pugh (Beeston).

BBJ 16 Aug 1894

The Derbyshire BKA had classes for honey, &c., at the Draycott and Wilne Floral and Horticultural Societies' show on August 4. Mr. Arthur G Pugh, Hon. Sec, NBKA, acted as judge.

BBJ 30 Aug 1894
A meeting of representatives of the Northern and Midland Affiliated Associations of the BBKA was held at Shrewsbury on Wednesday last in connection with the annual exhibition of the Shropshire Association. The meeting was a large and influential one, representatives being present from NBKA Mr. W. Broughton Carr and Mr. Jesse Garratt, members of the committee and the secretary of the BBKA, were also present.

The principal item on the agenda was "The consideration of the best means for giving effect to the recommendations of the Northern Association's Meeting, held in 1893, in respect to the formation of centres for conducting third-class examinations."

Considerable discussion followed as to the most suitable times and places for holding such examinations. Mr. Pugh (Notts.) considered that the second day of an exhibition was most suitable, as giving the candidates the opportunity of attending the exhibition and undergoing the examination at a moderate cost. Mr. Pugh moved "That it be recommended to the BBKA that where possible such examinations be held on the second day of a county show." Seconded by Mr. Meadows (Leicester), and carried.

Healthy 1894 Fertile Native Queens, 3s 6d each. Four Stocks same strain have this season yielded over 100 lb each. A. Simpson, Mansfield Woodhouse, Notts.

BBJ 13 Sep 1894
NBKA Bee and Honey Show at Moorgreen. The "combined" annual shows of the Eastwood and Greasley Agricultural and Horticultural Societies was held on September 4, at Moorgreen and the arrangements for the show of bees and honey were under the auspices of NBKA, for whom Mr. A Warner (local secretary of the NBKA) ably officiated as steward. The arrangements and staging for the honey department were of the highest order and reflected great credit on the management. The quality of the exhibits was good all round, extracted honey being especially so. Mr. Peter Scattergood, jun., of Stapleford, was the judge and made the following awards:
Twelve 1-lb. Sections— 1st, J. Annibal (sic); 2nd, H. Wiggett
Twelve 1-lb. Jars Extracted Honey— 1st, W. Lee; 2nd, G. Wood; 3rd, Mrs. White; hc .J. Rawson
Observatory Hive with Bees— 1st, J. Rawson; 2nd, H. Wiggett; 3rd, G. Reeve
Twelve 1-lb. Jars Extracted Honey (local class)— 1st, H. Wiggett; 2nd and 3rd, A. Warner and G. Reeve equal; 4th, JW Rawson
Frame of Honey in Comb - 1st, A. Warner; 2nd, J. F. Simpson; 3, J, Annable

BEEKEEPING IN VICTORIAN NOTTINGHAMSHIRE

J Annable was born in 1872 in Lenton. He was employed as a shorthand clerk. This is a classic example of the variable spelling of names which takes place in the records of NBKA.

Minute Book – Quarterly Committee Meeting was held on October 6th, 1894 at the People's Hall, Nottingham at 3pm. Present were Messrs Baguley, Faulconbridge, Forbes, Poxon, Rawson, Mr P Scattergood (Chairman), Warner, Hallam, Burton, Warner, Wilson, Wood, Fisher, Turner and AG Pugh (Hon. Secretary)

Mr Wood moved and Mr Hallam seconded that the minutes of the last committee meeting as now read be hereby confirmed.

Mr Scattergood proposed and Mr Baguley seconded that quarterly balance sheet showing a deficit of £2 0s 6d should be adopted.

The Secretary read the correspondence including letters of apology for non-attendance from Viscount St. Vincent and from Messrs Measures, Marriott and Finn also three letters from Mr Pett and a reply from Viscount St. Vincent with reference to minutes of last meeting.

Mr Scattergood moved and Mr Wood seconded that papers lie on the table and no further notice be taken of this matter.

A claim for loss of wax at Beeston Show, made by Mr Elliott, was read. Mr Fisher moved and Mr Burton seconded that Mr Elliott's attention be drawn to the latter part of Rule No. 6 which provides that the Society should not be responsible for such losses.

Mr G Wood had placed upon the Agenda a proposition that the present system of electing the Committee should be changed. After some discussion it was decided to leave the matter for further consideration at the next meeting.

A little discussion also took place respecting the provision of luncheon, refreshments, etc. for the stewards and others at shows and the advisability of distributing foul brood remedies amongst members. The secretary also said he should be pleased if anyone could make any suggestions towards making the approaching annual meeting more interesting and popular but no further minutes were passed.

The meeting closed at 5pm.

BBJ 18 Oct 1894
Meeting of the BBKA committee held at 105 Jermyn-street, on 11th inst. TW Cowan in the chair. On the motion of Mr. Scattergood, seconded by the Chairman, it was

resolved: "That the committee have heard with great regret of the death of Mr. FR Cheshire and desire to express their sympathy with Mrs. Cheshire in her great bereavement."

BBJ 18 Oct 1894
The nineteenth annual show of the British Dairy Farmers' Association took place at the Agricultural Hall, London, on the 9th inst. and three following days. The duties of judging devolved on Messrs W. Broughton Carr and JM Hooker, who made the following awards:
Twelve 1-lb Jars of Extracted Honey — c, W. Lee, Southwell

BBJ 18 Oct 1894
It may perhaps interest some of the readers of the Journal if I give my experience with foul brood this season. It may be of some little help to those who are unfortunately troubled with it, as I fear a good many bee-keepers are. Two or three years ago it existed about two miles away, which caused me to fear it might reach here. I have had several lots of driven bees yearly, but can form no idea where the disease came from.

The following are some notes of my treatment:
This spring I scarcely examined my bees, except to see they were not starving, until May 10, when I noticed two or three hives not working much. I had about twenty stocks standing near to each other, including three on the "Wells' system." I found No. 12 badly affected, so I burnt the bees, bars, brood, and quilts, and disinfected the hive.

I next sent off for some naphthalene and naphthol-beta. May 15, I examined Nos. 4 and 15, both of which looked suspicious by not working and found them bad. I, therefore, burnt the brood combs of each and fed No. 4 with carbolic, 11 drachms to 3 lb of food. No. 15 with naphthol-beta, 9 grains to 3 lb of food. This I poured on the empty cells. I then obtained some of McDougal's disinfectant carbolic and used about three tablespoonfuls to one pint of water and, with a paintbrush, used it on the frame ends, round the inside of hives and floorboards and for our hands, the smoker, or anything else employed amongst them. I still keep it at the hives, continually renewing when necessary. I then examined carefully all the hives, and found several more or less affected. I gave naphthol-beta to each in the food, and used naphthalene in the hives. May 25, I found in No. 4 some clean young brood. Again fed with carbolic, 11 drachms to 3 lb food. In No. 15 found some clean, healthy brood. Fed with 9 grains naphthol-beta to 3 lb food. May 28, fed again, but found one or two diseased cells.

On this date the bees were gathering scarcely anything, so went on feeding the lot—

half with 2 drachms carbolic to 3 lb food, pouring it on the combs; the other half, in the same way, I gave 9 grains naphthol-beta to 3 lb of food. Most of them were fed twice in this way. June 18, the expert and I examined Nos. 3, 4, and 15, and found them still affected. June 24, No. 3 swarmed; I took the bees and bars out of the hive and painted it over with strong carbolic, put in nine fresh bars of comb and one bar of fresh brood containing also a queen-cell; I then returned both bees and swarm, having found and destroyed old queen. Many of the bees joined No. 4, through the hive smelling so strongly of carbolic. They, however, reared a queen and I have found plenty of brood, but not any disease. June 24, I had most of my top boxes on, so could do no more doctoring. Most of them did fairly well, considering the scarcity of white clover here. August 5, took top box off No. 4. I thought it a little diseased below. Examined No. 3; it appeared to be quite clean.

I was recommended to use phenyle manufactured by Morris, Little, & Son, sold by most chemists. I have seen it also recommended in the Guide Book. I have also read Simmons's and Cheshire's books on foul brood. August 5, again examined No. 3; thought them quite clean. No. 4 I found a little diseased; I sprayed it with phenyle (13 drops to 8 oz water). I use it first on the bees and bars, then shake off the bees and spray thoroughly the bare combs. The bees are, with the spraying, almost as quiet as flies and I can recommend it to any one for manipulating with, except for taking honey, which it might flavour. During the season I have found two-thirds of my stock affected, but this autumn I have frequently sprayed and have punctured all foul cells I could find with the point of a stick continually dipped in a solution, half carbolic and half water.

September 1, I shook bees from combs of No. 14 into a straw skep for two days, then put clean boxes into a clean hive and poured on the combs 2 drachms carbolic to 3 lb food, then returned bees. So far, they appear clean.

I took the queen from No. 9 while it was affected. They reared another and she has bred all right; but again finding several affected cells I shook the bees into a skep for two days, then took away the queen, which was reared in the affected hive, united the bees with a driven lot, put on eight clean frames, and fed with medicated food: 2 drachms carbolic to 3 lb. food.

The other queen I kept for two days, and on September 20 joined her to another queenless hive, being desirous to know whether her brood would be affected or not. October 5, I examined these, and found a nice patch of sealed and unsealed brood, but could not detect any foul brood. I have been at a great deal of trouble, spraying with phenyle, washing the hives with carbolic, pouring medicated food on the combs and injecting carbolic into every sealed cell that I found without brood and I hope that I have now succeeded in getting them clean, although I must wait until spring to prove it.

What I would recommend to others finding a hive badly affected— burn all but the hive. The bees may be saved if shaken into an empty straw skep for two days, then put on clean combs and fed with medicated food. If one or two bars only are slightly affected, take them away. If one or two cells are found affected, puncture each cell with a match dipped in half carbolic and half water, then spray the rest with phenyle, 13 drops to 8 oz of water. In manipulating my bees I have treated each hive as though affected. J. Wilson, The Gardens, Langford Hall, Newark.

BBJ 8 Nov 1894
AE Trimmings (Notts.) The sample sent may be called pure bees' wax so far as not having been intentionally adulterated, but it nevertheless contains a considerable admixture of pollen which will account for the mealiness of texture, or grain, in the wax after melting.

BBJ 15 Nov 1894
The usual monthly BBKA Committee meeting was held at 105, Jermyn-street, on 8th inst. Mr. TW Cowan in the chair. The Secretary reported that, in accordance with the resolution passed at the last meeting, he had communicated with the several affiliated associations in reference to the formation of centres for conducting third-class examinations. The NBKA suggesting that the counties of Notts, Derby, and Leicester should be grouped together as one centre, the examination to be held on the second day of the exhibition held in either of these counties.

BBJ 29 Nov 1894
Destroying Weeds about Hives. Your querist, Ambrose A. Ogle (p. 458), last week anticipated a question I had been intending to ask you for some time. The "weed-killer" alluded to probably contains arsenic and I have used with great success a similar preparation on my garden-paths away from the hives. I believe carbolic acid, 1 oz to a gallon of water, to be an efficacious weed exterminator and should imagine that bees would not touch that owing to their objection to the smell. Should you see any objection to the use of the latter round the hives? Now I think of it, the proportions seem wrong and I should imagine it would be too weak to destroy the weeds. I have not yet tried it, but I had made a note of it from a gardening paper.
Percy Sharp, Brant Broughton, Newark-on- Trent, November 20.
Reply. There can be no objection to the use of carbolic acid about hives, no matter of what strength it may be applied. The bees will give it a wide berth if left to themselves and so no harm would follow.

BBJ 6 Dec 1894
Splendid Quality Honey. Two Tins, 70 lb in each; on rail at 7d per lb. Tins returnable.

BEEKEEPING IN VICTORIAN NOTTINGHAMSHIRE

400 Breffit's 1-lb Screw-Capped Jars, on rail at 8s 6d doz. A.
Simpson, Mansfield Woodhouse, Notts.

BBJ 3 Jan 1895

The examinations for BBKA second-class certificates were held at various centres on October 26 and 27, 1894. Ten candidates presented themselves for examination. Amongst the successful candidates was P. Scattergood.

Minute Book – The quarterly meeting held on January 5th, 1895, at the Peoples' Hall, Nottingham at 3pm. Present were Messrs GH Elliott (Chairman), P Scattergood, Faulconbridge, Forbes, Poxon, Rawson, Warner, Hallam, Turner, Burton, Wood, Marriott, Merryweather and the Hon Sec. (AG Pugh).

The minutes of the last meeting were read and confirmed. The quarterly balance sheet was read by the secretary. Proposed by Mr Scattergood and seconded by Mr Forbes that the honey labels in stock be considered as of value £2 10s and treated as an asset for that amount in the quarterly and annual balance sheets and that with this arrangement the quarterly balance sheet as now read be adopted.

The Hon. Secretary read the correspondence including apologies from Messrs Measures and Finn for non-attendance and a letter from the Secretary to the Hucknall Horticultural Society asking whether it could be arranged to hold a show at Hucknall in July similar to what has been done of late years. The committee instructed the Secretary to write accepting the invitation.

Proposed by Mr Scattergood and seconded by Mr G Wood that the annual meeting be held on Saturday, March 9th, 1895.

Proposed by Mr G Wood and seconded by Mr J Burton that the annual meeting be carried out on usual lines ie. election of officers in afternoon followed by a meat tea, after which Mr P Scattergood offer to read a paper on "Foul Brood" and exhibit a quantity of slides be accepted, to conclude with usual prize drawing. etc. It was decided to purchase from Mr R Turner, a honey ripener and from Mr Meadows a hive for the drawing. Mr Elliott offered to give a rapid feeder and Messrs Marriott and Pugh two sections each.

Proposed by Mr Merryweather and seconded by Mr Faulconbridge that the secretary be empowered to have the usual invitation circulars printed and distributed and that he be asked to write to Lord Belper, the Mayor (Joseph Bright), Town Clerk (Sir Samuel George Johnson) and Sheriff of Nottingham (William Evelyn Denison) to be present.

The secretary having explained that the distribution of the "Beekeepers' Record" is still very unsatisfactory, an animated discussion took place and it was finally decided that, in future, only those who pay 1/- a year in excess of their usual subscriptions shall have the "Record" and that those who pay such shilling extra have a copy sent them for their own use post free each month.

Henry Strutt, 2nd Lord Belper was born in 1840 and lived at Kingston Hall. He died in 1914.

BBJ 31 Jan 1895
The BBKA Educational Sub-Committee recommended the following as suitable centres for conducting third-class examinations during 1895 - Group 1. Notts., Derbyshire and Leicester.

The BBKA Special Foul Brood Committee had drawn up a list of statistics and other information which they required and had it inserted in the BBJ, in order to get the information from those in a position to give it. A marked copy of the Journal had been sent to Secretaries of County Associations, with a letter asking them to furnish the committee with information. Up to the present replies with information and suggestions had been received from Notts. amongst others.

BBJ 28 Feb 1895
The final meeting of the BBKA council elected for 1894 was held at 17, King William-street, Strand, on the 21st inst. Communication was received from P. Scattergood (Notts.) regretting his inability to be present. A letter was read from the hon. secretary of NBKA offering suggestions in reference to third-class examinations.

Minute Book – Annual General Meeting held at the Peoples Hall, Heathcote-street, Nottingham on Saturday, March 9th, 1895.

JL Francklin, Esq. who had kindly promised to preside in the unavoidable absence of Viscount St. Vincent who is away from England, wrote that a severe attack of influenza confined him to his house and prevented him from taking the chair much to his regret. Dr FR Mutch, Sheriff of Nottingham, kindly agreed to preside and the meeting commenced at about 3.15pm when there was a fair number of members present, a large number of apologies were read from absentees who were away on account of the influenza.

The secretary having read the minutes of the last meeting annual meeting, Mr Marriott proposed and Mr Wood seconded that the minutes of the last annual meeting as now read be confirmed.

The secretary read the annual report which was again of a satisfactory character for, although membership has decreased from 207 to 198, the amount of subscriptions received was the highest ever taken, the honey season was reported as being fairly good and the honey shows have been a success.

Mr Wood took exception to a remark in the report that it was a matter of regret that the annual show was only of such short duration that, in consequence, the exhibits which were of an exceptionally high standard could only be seen by a comparatively small number of members, he contending that a one day show was better than one that lasted two days from an exhibition point of view. Mr Hayes proposed and Mr Baguley seconded that the annual report as now read be adopted, printed and circulated to members.

The secretary then read the annual balance sheet comparing the same with the previous one and pointed out that, owing to the munificence of the Viscount St. Vincent, there was a small balance in the treasurer's hands at the year's end which was a new feature as in all previous years there was a deficiency. Mr Riley proposed and Mr Clark seconded that the balance sheet as now read be adopted.

Mr Pugh proposed and Mr Marriott seconded that our best thoughts be most heartily given to Viscount St. Vincent for his past services as President of this associationand that he be cordially invited to accept that position for the coming year. Mr Mills proposed and Mr Turner seconded that Mr AG Pugh be re-elected as Hon. Sec and treasurer on same terms as last year. Mr Warner proposed and Mr Baguley seconded that Mr P Scattergood be thanked for past services and be re-elected as Hon. Auditor.

Mr Wood proposed and Mr Newton seconded that members whose names are shewn in the report as acting as experts for the association be on the committee the same as district secretaries.

The election of committee members was then proceeded with and the following gentlemen were elected to serve during 1895: Messrs J Brooks, Elliott, Faulconbridge, Forbes, Marriott, Merryweather, Poxon, Puttergill, Turner, Webster and White. Messrs Pugh and Scattergood were re-elected as joint representatives to meetings of BBKA. A vote of thanks was awarded to Dr Mutch for presiding and after he had responded, Messrs Meadows and Wood judged the honey that had been sent for tea and awarded certificates for extracted and comb honey to Mr Wiggett of Hucknall.

At 4.40pm about 50 members partook of tea and at 6.30pm re-assembled for the conversazione at which Mr Scattergood had promised to read a paper on foul brood; he was, however, unable to attend owing to illness. Mr Craven was elected chairman and having read some extracts from daily papers re 'a lecture given at West Bridgford' called on the secretary to distribute medals, etc. which were awarded as follows: BBKA Silver medal for best extracted honey at Southwell Show Mr White, Newark

BBKA Bronze medal for best comb honey at Southwell Show Mr Wiggett, Hucknall
BBKA certificate for best granulated honey at Southwell Show Mr Merryweather, Southwell

Some discussion took place with reference to the annual show and it was decided to ask the committee to offer prizes for mead, vinegar and other articles in whose composition honey forms a principal part, for hives made by amateur and also to offer a substantial prize for best collection of appliances. Mr Raven said he did not think prizes for wax were large enough and said he would make a gift of 10/- per first prize in that class, present prizes to go to 2nd and 3rd best exhibits.

It was also decided that wherever the annual show is held, if the local society do not invite stewards, judge and secretary to luncheon the same be provided from the funds of NBKA. Also that during 1895 the winner of a first prize in honey classes at local shows, be debarred from showing in such a class at other shows during the season.

The prize drawing then took place and the meeting ended about 9pm. The following prizes were won:

"Wells" hive	won by Mr. GE. Puttergill
"XL all" hive	won by Mrs. Hallam
do. do	won by Mr. W. Herrod
honey ripener	won by Mr. Wright
rapid feeder	won by Mr. Annable
Beeston feeder	won by Mr. Wiggett
100 sections	won by Mr. Breward
100 sections	won by Mr. Measures
volume of Record	won by Mr. Fox
a "perfect smoker"	won by Mr. Brooks
an improved super clearer	won by Mr. Draper

Major John Liell Francklin b. 1844, d. 1915. He married Hon. Alice Maude Jervis, daughter of Carnegie Robert John Jervis, 3rd Viscount St. Vincent of Meaford and Lucy Charlotte Baskervyle-Glegg, in 1868. He lived at Gonalston Hall, Nottinghamshire
Dr Forbes R Mutch, General Practioner, Sheriff of Nottingham 1894/5, was born in Scotland in 1856.

W Herrod is, of course one of the brothers (William (1873 – 1951) and Joseph (1870 – 1958)) born in Low-street, Sutton-on-Trent. They attended the local inter-denominational Board School. William claimed that their parents Thomas and Ellen, intended them to become elementary school teachers but they followed in their father's footsteps as carpenters.

BEEKEEPING IN VICTORIAN NOTTINGHAMSHIRE

They started beekeeping in 1888 as a hobby, making their own hives and were still living at home according to the 1891 census. They were taking part in NBKA shows as J and W Herrod, Trentside Apiary during the last decade of the century. During the period discussed here they were frequent competitors in the various honey shows either singly or as a pair.

In 1894 William, having devoted all his spare time to beekeeping, went to work at the John Howard bee appliance manufactory near Peterborough. In 1896 he was appointed Touring Expert to the Lancashire BKA where he organised their exhibit for the Royal Show in Manchester. The following year the BBKA appointed him as National Expert based at the Swanley Horticultural College, in Kent.

Joseph qualified as expert, 3rd class, in 1899 but I cannot find out at this time where and when William qualified as an "expert". In 1898 Derbys BKA refused William's application to be a candidate for the BBKA examination unless he became a member of that organisation. However, he was advertising himself as "expert of the BBKA at Swanley" by 1899 and he was also looking for candidates to coach for the BBKA examinations at this time.

Joseph Herrod

William Herrod

BBJ 23 Mar 1895
Under the joint auspices of NBKA and the Notts. County Council a lecture on "Beekeeping" was given in the Board School, Keyworth, on Wednesday evening, by Mr. Arthur G Pugh, hon. secretary of the association and certified expert of the BBKA. There was a large attendance and the chair was occupied by the Rev Robinson. The lecturer dealt with the anatomical and physiological structure of the bees, the various diseases of bees, the different kinds of hives and apiaries and the various methods of managing bees. All the subjects were fully illustrated. The object of the lecture was to stir up an interest in beekeeping as a profitable industry. The lecturer stated that during 1894 no less than £33,475 worth of foreign honey was imported into this country, thus showing that there is a demand for the product. For an outlay of 15s on a hive and another 15s on bees and accessories, a return of 15s per annum could be secured. On the motion of Mr. Neale a hearty vote of thanks was accorded to Mr. Pugh for his lecture.

Rev. Harold Robinson BA was born in Leicester in 1869. He was vicar of Plumtree from 1926 until 1936. He died in Eakring in 1944 and was buried in Coleorton. He was described as a big jovial man with a hearty laugh and a friendly and welcoming manner.

NEP 16 Mar 1895

Under the auspices of the Technical Education Committee of the Council, a lecture on bee-keeping was given in Oxton Church Schoolroom, by Mr. Arthur G Pugh, the honorary secretary of the NBKA. Mr. Pugh admirably described the three classes of the domestic bee, as their form, their functions, and their habits and illustrated their description with a series of excellent magic-lantern views, and by a quantity of bee-keeping apparatus lent for the purpose by Mr. George Wood, an Oxton beekeeper, who gained the silver medal last year from the NBKA for honey out of the comb. At the close of the lecture, the Rev. Prebendary ER Mason, vicar, who presided, moved a vote of thanks to Mr. Pugh.

NG 16 Mar 1895

The annual general meeting of the NBKA was held in the People's Hall, Heathcote-street, Nottingham, on Saturday afternoon. Mr. JL Francklin, JP. was announced to take the chair, but he was unable to be present owing to an attack of influenza and his place was taken by Councillor FR Mutch, MD. (Sheriff of Nottingham). There was a goodly attendance, amongst those present being Messrs. Wood, AG Pugh, Riley, Merryweather, Poxon, Hayes, WP Meadows, Faulconbridge, Newton, Marriott, Mills, Turner, Radcliffe, Wootton, Warner, Maskery, Forbes, Baguley, Burton, Burrows, and White. A number of letters were read from members unable to be present through indisposition.

The annual report stated that the membership showed a slight decrease, 42 members having ceased to subscribe, whilst 33 new members had been enrolled, making a net loss of nine, the totals being 207 at the commencement, and 198 at the close of the year. It was, however, pleasing to state that there was no loss in amount of subscriptions. For the first time since this association commenced active work there was a balance in the hands of the treasurer, which was due to the munificence of the president. The amount of honey produced in the county was reported to have been about the same as in previous' years, that was to say, rather above the average and of a good quality. According to a resolution passed at the last annual meeting a honey label was adopted and a supply promised, of which members had purchased 4,400 and they appeared to be much appreciated by both members and their customers.

Foul brood was unfortunately on the increase, although it was not of so virulent a type as reported in some counties. Special measures were being considered by the BBKA committee to prevent its spread and it was earnestly hoped that every member would take every possible precaution to prevent the spread of this pest. Experts had visited all members' apiaries and their reports were of a satisfactory character.

The Notts. County Council, having so many calls on their funds had unfortunately had to reduce their grant in aid of technical education in bee-keeping from £40 to £20 per annum. The balance-sheet showed a small balance in hand. Both reports were unanimously adopted.

On the motion of Mr. AG Pugh, seconded by Mr. Marriott and supported by several members, Viscount St Vincent was unanimously re-elected president for the ensuing year. Stress was made upon the fact that his Lordship had done a great deal towards the success of the association and of bee-keeping in general. Mr. Mills proposed and Mr. Turner seconded that Mr. Pugh be re-elected hon. secretary, both gentlemen speaking highly of the qualifications and of the energy of Mr. Pugh. The resolution was heartily carried. Mr. Scattergood was re-appointed auditor and a committee was appointed. The proceedings were followed by a tea and conversazione during which the usual drawing for prizes, presented by the president and various other donors, took place.

Leicester Chronicle 30 Mar 1895
The annual meeting of the Leicester BKA was held on Saturday in the Old Town Hall. There were over thirty persons present, including several ladies. At the outset it was intimated by Mr. HM. Riley that Sir Israel Hart was unable to be present, having had to leave Leicester to visit a sick relative, and Sir Thomas Wright was also unable to attend owing to illness. Mr. Pugh, secretary to the Nottinghamshire BKA, was, in the absence of the gentlemen above-named, unanimously elected to the chair.

Minute Book – quarterly committee meeting held at the Peoples' Hall, Nottingham on April 6th, 1895 at 3pm. Present Mr GJ Elliott (Chairman), Messrs J Brooks, Faulconbridge, Forbes, Marriott, Merryweather, Poxon, Puttergill, Hallam, Measures, Scattergood, Pugh and Warner.

The minutes of the last quarterly meeting were read and confirmed. The minutes of the annual meeting were also read and discussed. Correspondence was read from the president, Mr JB Cluro, Mr HJ Raven, Mr Geo Wood, Mr J Holmes, Mr J Huckle and the secretaries of Lowdham and Burton Joyce Horticultural Societies. The secretary was instructed how to deal with the same.

NEP 17 Apr 1895
Loan of the Shire Hall. The Chairman, upon the reading of letter from Cllr. Allsebrook, to the loan of the Shire Hall for meetings of the Beekeepers' Association, said that he declined permission on the first consideration of the matter and remarked if the committee granted the application to this association they could scarcely refuse

the applications that might be made on behalf of other associations. It was for the committee to decide. Mr. Mellors moved that permission be granted. The resolution was not seconded and an amendment, moved by the Chairman and seconded by Mr. FE Seely, that permission be not granted, was carried.

Minute Book – Special Committee Meeting held at the Peoples' Hall, Heathcote-street, Nottingham at 3pm on Saturday April 27th, 1895. Present were Messrs Mackender (Chairman), FT Faulconbridge, SW Marriott, W Poxon, GE Puttergill, G Webster, AE Newton, J Rawson, Leath and the Hon. Secretary. Letters were read from the President and Messrs Measures, Wood, and Holmes regretting their inability to attend.

An application was read from the Norwell Horticultural Society asking for assistance in Bee and Honey Dept. at their show on August 12th 1895. It was decided to grant at the same terms as are in place for Lowdham.

A formal application or invitation was received from the Newark Pleasant Sunday afternoon association asking NBKA to hold their annual show in conjunction with theirs on Bank Holiday, August 5th 1895 and offering to provide tent, staging, etc. and give £7 10s towards prize list and agree NBKA taking the whole of the entry fees. Mr Newton moved and Mr Puttergill seconded that the invitation be accepted and the annual show to be held accordingly. The various clauses of the annual schedule were then discussed and the secretary instructed how to deal with the same. Mr EN White was selected as judge and Mr SW Marriott agreed to act as his assistant.

Robert Mackender was for many years a NBKA committee member He was born in Mildenhall, Suffolk in 1849 and was married in Bromley, Kent in 1874. He died in Newark in 1932. He was for many years head gardener at Staunton Hall, near Newark. He then carried on a seeds and bee appliance business in Newark for 40 years.

He acted as mentor to the young Herrod Brothers in their early days.

BBJ 2 May 1895
AE Furnival (Notts.) Comb received is probably from a hive the bees of which perished from foul brood, but so long ago that all traces of the diseased larvae have been dried up and disappeared, only to be traced may be by much trouble and the use of the microscope. No need to destroy the hives if thoroughly disinfected.

BBJ 16 May 1895
Several Healthy 1894 Queens for Sale, 3s 6d each. A. Simpson, Mansfield, Woodhouse.

WH. (Newark) Queen received has never been mated.

Are Young Fertilised Queens Left in Hives when First Swarms Issue?
1 Do you think it ever occurs that a young fertilised queen is left in the hive when the old queen leaves with the first swarm?
2. If queen, as above, was in the hive, would she allow four or five queen-cells to remain untouched for two days?
3. If such a queen was at work would the bees fill the centre of combs with honey — said combs being in the middle of brood-nest?
4. With abundance of brood and some eggs, what is likely to be the outcome?
GD, Long Eaton June 12.
Reply
1. Pre-supposing that by "fertilised" a laying queen is meant, we have never heard of such a thing and consider it very improbable in your case.
2. A recently hatched queen, whether fertilised or not, will not tolerate embryo rivals in the same hive and unless the latter are protected by the worker bees, they would soon be destroyed.
3. The bees would fill with honey any cells found vacant, whether in the centre or not, so long as no egg-laying was going on in the hive.
4. When the young queen left in the hive begins ovipositing, all will, no doubt, go on right.

BBJ 13 Jun 1895
August 5 (Bank Holiday). NBKA Annual County Show on the Sconce Hills, Newark. Eighteen classes for bees, hives, and honey, with liberal prizes. Several open classes. Entries close July 27. AG Pugh, hon. sec, NBKA, 51, Mona-street, Beeston.

BBJ 27 Jun 1895
Royal Agricultural Society Show at Darlington.
The weather on Monday was brilliantly fine the whole day, and the show week promises to be one of the most successful on record.
Most Complete Frame-Hive. hc, Geo. E Puttergill, Mona-street, Beeston

Minute Book – Quarterly Committee Meeting held at the Peoples' Hall, Nottingham on July 6th, 1895. Present Mr P Scattergood, Jun., (Chairman), Messrs Brooks, Faulconbridge, Marriott, Puttergill, Rawson, Warner, Newton, AG Pugh and Newton.

The minutes of the last quarterly meeting and special meeting were read and confirmed.

The quarterly balance sheet was read. Mr Wood proposed and Mr Marriott seconded that the balance sheet as now read be adopted.

The secretary read letters from Viscount St. Vincent and Mr Measures regretting their inability to attend. Letters from Mr Walker, hon, secretary, Derbys. BKA, was also read re. "expert's examination" and it was decided to leave this matter in the hands of the secretary. A letter was also read from Mr Mortimore and the secretary instructed to advise him that no attendance could be made in the 1895 schedule.

Mr Scattergood was appointed judge for Southwell and AG Pugh for Moorgreen shows at a fee of 5/- plus travelling expenses.

The hon. Secretary (AG Pugh) having tendered his resignation of his present appointment at the end of 1895 which was very reluctantly accepted and it was decided to further consider the matter and nominate a successor at the next quarterly committee meeting so that the new officer can be getting initiated into duties.

The chairman complemented Mr Puttergill upon his getting a HC in the hive class at the Royal Show in Darlington and the meeting terminated at 4.20pm.

NG 6 Jul 1895
We have had an exceptionally dry time and though a few thunder showers have fallen, still there has been no weight of rain since the early part of May and vegetation began to look bare and dry. Notwithstanding this fact, in some districts round Nottingham honey is coming in rapidly and sections and supers are being nicely filled. A beekeeper of our acquaintance a fortnight ago removed 50 sections from two hives and has about 30 more nearly ready on two other hives, while a beekeeper about four miles from Nottingham on June 25th extracted 28 lb from supers on two hives. A beekeeper the other day asked me how he was to keep up the working impulse of his bees at high pressure speed and as this may not be generally known we give this hint in our corner for the benefit of our readers. By a little judicious management under certain circumstances this is not a difficult thing to do. For instance, when removing full boxes of sealed combs, if we can replace them with similar boxes furnished with ready-built combs and among these latter are placed two or three "wet" frames just from the extractor and dripping with honey, the bees will "tear away" with redoubled energy to re-fill them, working much harder than if dry combs or only sheets of foundation are given. Try it.

NG 6 Jul 1895
JG, Long Eaton, asks how to make a hive of vicious bees quiet, so that he can manipulate, or, as he says, take the hive to pieces without being stung. The smoker has very little effect.

In reply we must say there are several points to be attended to in quieting bees. First, do all your manipulating carefully. Bees dislike all quick movements about the hives, especially any which jar the combs, so do not get into a bustle and bang the hive or frames about and, if you do happen to get stung do not lose your head. Second, do not uncover all your frames at once and, above all, do not be constantly meddling with them and every day upsetting them.

If your smoker does not have the desired effect, try the following. Make a solution of $3/4$oz Calvert's No. 5 carbolic acid, $1/2$oz glycerine and one pint of warm water. Mix the acid and glycerine well together before the water is added. When cold, apply to entrance and along the top of the frames, with a small brush or a goose quill or spray diffuser. Another method and a good one. too, is to get a square of calico, about 18 or 20 inches square, and thoroughly saturate the same in the solution above-named, then wring out, fairly dry and lay the square on the top of the frames and unroll as you wish to manipulate. We have quieted some vicious stocks in this way, without being stung and last week took from a hive which was the *bete noire* of the apiary a crate of 21 sections without an attempt to sting.

NG 13 Jul 1895
WFB, Nottingham, writes: I am the possessor of a hive of bees, housed in two straw skeps one placed on the top of the other. Knowing nothing about bees, I followed the advice of a friend who suggested that in order to keep the bees from swarming last year that a second hive should be added in the form of a super. Now I want to place the bees in a frame hive, but since there is brood in both the skeps I find myself in a difficulty. What is the best method to adopt?
Reply
The best thing for you to do now is to place the frame hive in the position of the straw hives and put in full sheets of good foundation, say six or eight, then place the skeps on the top of the frames and be sure and cover all up snug and tight and place the hive cover of the roof over all to deter the wind. In a short time the bees will work out the foundation and the queen will descend and begin to lay in the new cells. Then in about eight or ten days look out for the queen in the frames in the lower compartment and if you discover her, then place a queen excluder (a sheet of perforated zinc, with perforations made the correct size, which may be obtained from appliance dealers, 16in, or 18in square for 1s 4d) on the top of the frames, this will keep the queen below and in about a month afterwards remove the skep from the top of the frames as all the brood will have hatched out above and joined the bees and queen below.

Then remove your skeps, and cover up the frames with your quilts and if, say by the middle of September on examination, you discover that the bees have not a good supply of stores, feed them with syrup. You can cut away the combs from both skeps and run or drain through a hair sieve the honey which may be in the combs after cutting them away.

I would advise yon to join the Notts. BKA and I have no doubt any one of their experts would do this little job for you in half an hour.

BBJ 18 Jul 1895
The BBKA Council had under consideration numerous communications in respect to the appointment of judges and examiners from the counties of Wilts, Yorkshire, Ely, Lincoln, Derbyshire, and Staffs. A letter was read from Mr. Scattergood, of Stapleford, Notts, suggesting that some arrangement should be made for endorsing the certificates of those candidates who had passed successfully in the "knowledge and treatment of foul brood." Resolved, "That the secretary do endorse all such certificates on their being returned to him."

NG 20 Jul 1895
The other day the general secretary of the NBKA sent me a schedule of the shows to be held by the association in different parts of the county in connection with the Agricultural and Horticultural Societies of the following districts: Lowdham, Hucknall, Southwell. Norwell, Newark, Moorgreen, and all the members are allowed to show at reduced entry fees. Substantial money prizes are offered, while at the annual show at Newark on Bank Holiday, silver and bronze medals, and pendants, etc., are offered in addition. The members have evidently plenty of advantages for their small subscription. Those who intend showing at any of the honey shows should get their honey bottled at once, so that it may clear and settle. It will be all the better if run through a piece of fine muslin and allowed to stand in the bottles for a few weeks.

NG 3 Aug 1895
The other day I went to the Hucknall Torkard Flower Show, where the members of the NBKA were having a honey show. Some good samples of honey were staged, and the honey shown by Mr. J Hind of Papplewick Grange, was of splendid quality and colour and deservedly secured the first prize. The County Council were also represented in the bee department and useful and interesting instruction on bees and their management was given in the lectures and talks in the bee tent.

NG 10 Aug 1895
The chief Bank Holiday attraction at Newark this year was the festival of the local PSA Society. The NBKA Show was a greater success than ever. This society holds several small shows during the year, but this is the premier exhibition. Mr. AG Pugh, of Beeston, the secretary, worked hard with getting entries, although the year has not been a good one for honey in Notts. The entries sent in were in excess of any previous show. Amongst the exhibitors were Viscount St. Vincent, Mr. Wiggett (Hucknall), and other well-known enthusiasts in beekeeping. Over half a ton of honey was sent in in pound jars and some of the entries were of a perfect colour and consistency. Viscount St. Vincent got the first award for the best trophy of honey, which was excellently

arranged, Mr. G Marshail (Norwell) coming second and Mr. W Lee (Southwell) third. In the amateur class for six best bottles Mr Herrod (North Clifton) came second. A couple of novelties in the shape of honey vinegar and honey cake were included in the schedule. Mr. Wilson (Langford) took the former prize with a nice bottle. The vinegar is a bye-product and is obtained from what has hitherto been an offal in beekeeping. The honey cakes were excellent and to each was attached a recipe showing what it was made of. Mrs. Fromings (Bridgford) was placed first and Mrs. Puttergill (Beeston) second. Mr. Merryweather (Southwell) sent the best observatory hive and the best amateur-made hive was sent in by Mr.WF Faulconbridge (Bulwell). Several lectures and exhibitions of bee-driving were given by members of the society to large audiences.

Sarah Fromings was born in 1863 in Carlton. Her husband, Thomas A, was also involved in the activities of NBKA. He was a county court clerk and was born in 1864 in Whittingham, Derbyshire.

NG 10 Aug 1895
GS (Carlton) says he should like to commence beekeeping and asks our advice.
It would be hardly be for us to suggest any line of action for you to follow. We might give you one word of caution, do not commence on too large a scale, but begin in a small way at first, then increase your stock as you gain knowledge and experience.

A Worker (Long Eaton) writes and asks:
(1) Would it not be best to leave the honey on the hive to ripen say till the first week in August, because he thinks his honey would be of superior quality?
(2) What advantage is there in frequent extracting?
(3) Should he under any circumstances extract from the brood nest?
(4) Would it be advisable to place the extracted combs in a convenient corner of the garden to be cleaned out by the bees?
To take the questions in turn:
1. Honey ripens best on the hive and it is certainly better in quality and much superior to that extracted when it is not fully ripe, but if you will avoid extracting till the honey has been sealed over for a few days you need have nothing to fear and especially if you can place it in a deep vessel and run it out or draw it off the bottom.
2. The advantage of frequent extracting is that as a rule more honey will be extracted if the honey is extracted once a week in the height of the season, the bees working with a will to fill the combs again.
2. We certainly do not advise you to extract from the brood nest, unless we have a honey glut and the bees fill the brood combs with honey and there is no room for the queen to deposit her eggs. Then, of course, you must provide more room for the queen, but in this case we would prefer to remove a frame or two of honey and place empty combs or sheets of foundation in the centre of the brood rather than extract the combs. Whatever the bees give you above be satisfied with and do not rob them

of what they may have below.

4. If you follow out the plan of leaving extracted combs in a convenient place to be cleared by the bees, the probability that your whole apiary and your neighbours also, if there is one, will be devastated and robbing and fighting will result. If you have a number of combs to clean you might make one colony clear out the whole lot by using a super clearer with a Porter Bee Escape in it.

NG 17 Aug 1895

The principal event among bee-men during the past week has been the annual show of the NBKA at Newark on Bank Holiday and it is certainly one of the most successful ever held by the association. There were 128 entries and nearly all staged and the honey was of a uniform standard of excellence. Mr. Meadows, of Syston showed a very good collection of appliances and Mr. GE Puttergill, of Beeston, deservedly won the first prize for the best and most complete and inexpensive frame hive. There was also a new departure, prizes being offered for the best bottle of honey vinegar. This may sound strange to our readers, but those who have used honey vinegar will not want to use the other kind again. A few subscriptions were given to the prize fund by the Mayor of Newark and others, and it is to be hoped that the example thus set may be followed by others who may be in a position to help and encourage the association.

Special attention may also be drawn to the fact that all the local honey staged was labelled with the new registered label of the NBKA and all honey sold bearing this label must be guaranteed pure by the seller. Thus the purchaser has the advantage of securing a pure article and not some of the vile compounds which sometimes are put up and sold as honey.

The honey harvest is now nearly over and as we advised last week it will pay to gently stimulate all hives, giving about a gill of syrup every night for the next three weeks. The syrup should be given regularly and slowly.

During the last week we paid a visit to several beekeepers in North Notts. In several villages we were shown fields white with the second crop of white clover and the bees were hard at work carrying in stores to the hive and thus saving the beekeeper's pocket and the trouble of feeding; lucky beekeepers we thought.

BBJ 20 Aug 1895

Do Bees Carry Eggs? Mr. Brice, in his "Jottings" (p. 292), refers to myself in connection with this interesting subject and, with your permission, I would like to detail the facts of the case, which to me are so full of interest.

Before doing so, however, let me correct an error Mr. B. makes in referring to my expert's certificate. At present I am only a second class expert, not "first," as stated. I may say that all my observations and examinations have been carefully made and, being noted at the time, the details now given I know to be correct. I received from Mr. Brice the queen referred to and a few workers safely in one of his own cages, on July 12 last.

I made the stock queenless a few days beforehand and had carefully examined the hive on the previous day. I also very closely inspected every frame about six o'clock on the evening the queen arrived, just before introducing her and there was no sign of eggs, to say nothing of unsealed brood in the hive. I followed the instructions sent with the queen, pressing the cage well into the comb and honey, I then left the hive alone till Monday the 13th, being of opinion that many failures in queen-introduction are due to too much fussy manipulation. However, I opened the hive on the date named and found the queen was still imprisoned, the bees not having liberated her by eating their way into the cage, allowing the queen to come out. I then took out the frame next to the one on which the queen was caged and found a few eggs deposited in worker cells. This made me wonder if a fertile worker was present in the hive. However, I let matters remain as they were and did not liberate the queen, nor did I again examine till Wednesday morning the 17th, when I found the queen still caged.

On examining the comb as before I found quite a number of cells (nearly 200) with young brood a few hours old and eggs, not deposited here and there all over a comb as a fertile worker usually deposits them, but quite in regular manner, in rows (on one comb only) four cells deep, having eggs a few cases near the centre of the patch five cells deep and only in one case did I see more than one egg in a cell. I then released the queen and she was at once joyfully accepted. Moreover, she is now laying well after having been caged nearly five days. Some may say "too long." Well, perhaps under ordinary circumstances it would have, but these were not ordinary circumstances and therefore I was in no hurry, if I could by careful observation learn something more about the bees. The three questions raised by Mr. Brice are so satisfactorily answered by him, that I need say nothing respecting them, save that his conclusions entirely coincide with my own; but someone may ask: " What has been the result?" and though, as Mr. W Woodley points out (p. 203), it is one of those inexplicable things which occasionally happens.

The important fact, however, remains, that the eggs deposited while the queen was caged have produced both workers and drones and the latter were much lighter and brighter in colour than any in my apiary, my bees being all blacks. The fact of workers and drones of a distinctly lighter colour to any of my bees resulting from the eggs deposited in the cells as noted above, furnishes a complete corroboration of the theory that the eggs were laid by the queen while caged, and were carried by the bees into the cells in which the workers and drones mentioned have been reared. But for trespassing too much on your space, I could give the days and dates of my notes and observations, but probably sufficient has been said on the point, except to add how pleasing it would be if some of our scientific bee men would give us their opinion on this, to me, interesting part of our study.

Peter Scattergood, Stapleford, Notts, August 19.

NG 31 Aug 1895

A few years ago Mr. WPJ Allsebrook of Wollaton discovered in a hedge of one his fields in the late autumn time the work and remains of a swarm that had so absconded and it contained several combs, beautifully built.

NEP 3 Sep 1895

Moorgreen Agricultural and Horticultural Show. Brilliant weather favoured the inauguration of the seventh annual combined exhibition at Moorgreen this morning, in connection with the Greasley, Selston, and Eastwood Agricultural and Horticultural Societies, with Major Rolleston, JP. Vice-president. The stages in a tent occupied by NBKA were well filled, the exhibits for the season being of a high quality and the judge, Mr. P Scattergood, jun., experienced some difficulty in deciding upon the merits of each.

BBJ 3 Sep 1895

A New Feeder. In answer to Mr. WB Webster's remarks, (p. 383), permit me to explain my method of feeding, which, I think, can be utilised for a rapid feeder, and yet do away with the bits of muslin, and your correspondent's two days' work; it also carries out his idea of a feeder close to the frames.

I use a wood cover with $1/4$ in. space above frames; this gives ready access to every frame. In centre of cover I cut - with a fret-saw - a hole $3^{1}/_{8}$ in. across. I then take two ordinary self-opening 2-lb. golden-syrup tins; the lids I fit with a piece of $1/2$ in. wood cut as follows - one half of the circle is made $3^{1}/_{8}$ in. across, the other half is $2^{5}/_{8}$ in. across, this leaves $1/2$ in. space half-way round the lid.

My next step is to fasten the wood into the hollow of the lid with a screw in the centre from inside. The wood will now turn round like a wheel, and shut off to any number of holes I require; a dozen holes half round the lid completes the feeder, which holds just 1 lb of sugar when made into syrup. I fill my tins at home, put on the lid, turn the wheel and close the holes for carrying it to the bees.

Arrived there, the hive is uncovered and the wheel of tin turned to the number of holes required; give the empty tin a screw-round to release it from propolis, lift it off rapidly and replace with the full one. The few bees adhering to the empty feeder-lid are allowed to run in at the hive entrance.

I have been well pleased with this feeder. It works well in practice. Fifty tins can be filled if required, and charged, the lids changed as wanted for use. There is no trouble at the hive. The merest novice, if veiled and a carbolic cloth ready, need not fear to change the tins and be safe against accident. As the wood stands up above the lid and the bees cluster on it, very little tin comes in contact with the bees; besides, a tin vessel is also safer against accident than glass.

Here is Mr. Webster's rapid feeder at once: two self-opening 7-lb. tins (larger if you like), fitted-up as above, will just meet our friend's wants. Where a quilt is used instead of a wood cover, a stage will have to be made of $1/2$-in. wood with a fret saw, and a $1/4$-in. window lath nailed on for bee-space.

A Worker, Long Eaton, September 31.

BBJ 5 Sep 1895

The annual show of NBKA was held at Newark on Bank Holiday August 5th, in weather eminently satisfactory. The exhibits were staged in a spacious tent and presented a most pleasing and effective attraction to the show in connection with which the exhibition was held. Nearly the whole of the 128 entries were staged, and, according to the general opinion, formed the largest and best show the Association has yet held. The open class brought together some exceptionally good honey, while the local classes, considering the very moderate season they have experienced, did the members great credit. In the class for display of appliances Mr. Meadows' exhibit was extensive and meritorious.

To Mr. Pugh, the energetic hon. sec, the thanks of members are ungrudgingly given, as also for the ready assistance of the stewards. The judge, Mr. CN White, of Somersham, Hunts, was assisted by Mr. SW Marriott. During the day Mr. White delivered a series of lectures in the bee-tent and also conducted an examination for 3rd class certificates.

Best Hive—1st, GE Puttergill, Beeston
Amateur-made Hive—1st, JT Faulconbridge, Bulwell; CH Picken, Fledborough
Trophy of Honey — 1st, Viscount St. Vincent, Norton Disney; 2nd, G. Marshall, Norwell; 3rd, W. Lee, Southwell
Twelve 1-lb. Jars Extracted Honey—1st, AJ Mortimer, Oxton; 2nd, R. Mackender, Newark; 3rd, H. Merryweather, Southwell; 4th, JW Herrod, N. Clifton *(sic – should this be J&W Herrod if so why does William also compete on his own in the same event?)*; 5th, W. Measures, Upton; hc, J. Wilson; c, Viscount St. Vincent
Six 1-lb. Sections—1st, Viscount St. Vincent; 2nd, CH Pickin; 3rd, AJ Mortimer
Six 1-lb. Jars Granulated Honey—1st, H. Wiggett, Hucknall; 2nd, GE Puttergill; 3rd, H. Merryweather

Shallow Frame of Honey—1st, Geo. Marshall; 2nd, JW Herrod; 3rd, GE Puttergill
Six 1-lb Jars Extracted Honey (Amateurs) — 1st, CH Pickin; 2nd, W. Herrod; 3rd, GE Puttergill; 4th, T. Marshall, Sutton-on- Trent
Twelve 1-lb. Sections (Blow's) — 1st, CH Pickin
Twelve 1-lb. Jars Honey (Blow's jars) — 1st, CH Pickin
Honey Vinegar—1st, Mr. Wilson, Newark; 2nd, GE Puttergill
Honey Cake—1st, Mr. Fromings, Radcliffe-on-Trent; 2nd, GE Puttergill
Observatory Hive—1st, H. Merryweather; 2nd. W. Herrod; 3rd, T. Marshall; 4th, T. Maskery, Kirkby
Beeswax—1st, J. Wilson, Langford; 2nd, J & W Herrod, Sutton-on-Trent
Bee-driving—1st, W. Herrod; 2nd, AG Pugh, Beeston

NG 7 Sep 1895
Moorgreen Agricultural and Horticultural Show. This show was held on Tuesday under the most favourable auspices. The stages in a tent occupied by the NBKA were well filled, the exhibits for the season being of a high quality and the judge. Mr. P. Scattergood, jun., experiencing some difficulty in deciding upon the merits of each. Another interesting feature was an exhibition of butter-making in connection with the Notts. County Council Technical Instruction classes, Mr. WF Allsebrook being the judge. Another gentleman who kindly fulfilled the onerous duty of judge was Mr. A Warner of the NBKA.

NG 14 Sep 1895
A Worker (Long Eaton) asks do we recommend syrup in a vessel and letting the bees help themselves.

Reply. Yes, if the vessel is a feeder and placed on the top of the hive; but not a vessel left outside and the bees allowed to help themselves. A feeder, suitable for putting on the top of frame or even on straw skeps, if it can be securely fastened or covered up so that the bees of other hives cannot get at it. For stimulative feeding this is the best we know. It is called the Universal or Nottingham First Prize Bottle Feeder, having obtained the first prize at the Royal Agricultural Show held at Nottingham. It consists of a bottle with a metal cover in which are pierced a number of holes. The bottle and cap are placed on a block or stage in which is cut a long narrow semi-circular slot, through which the holes in the cap are visible and the index is so placed that any number of holes may be open to the bees. Personally the writer always gives the bees the full number of holes and lets them take down the syrup while just warm as quickly as they can.

For rapid feeding which should now shortly be resorted to, so that the bees may have an abundance stores and be sealed over by October 1st, the use of a rapid feeder is advisable. About six pounds of syrup may be given at once and if given just warm in the evening the bees will take it down in a night. This may be renewed till the bees have about 20 lb of sealed stores. Both these feeders described are made by Mr. Meadows, of Syston.

A Worker also asks is there any other way of giving candy for wintering than on the tops of the frames. Yes, by inserting a frame full of candy by the side of the cluster, but as it has serious disadvantages we do not favour this method. If, however, you want to place candy on the hive, put it on the top of the frames and cover all up snug and warm. Then the bees can get at it much better than if it is placed at the side of the cluster.

The other day we paid a visit to the Moorgreen Agricultural and Horticultural Societies Show, and were greatly pleased with the interesting exhibition of bees and honey. The tent in which they were staged attracted a large number of visitors during the day. The honey was of very good quality for the season, while the staging and arrangements of the exhibits evinced great taste and judgment on the part of Mr. A. Warner, who, as usual, officiated as steward of this branch of the show. A very interesting exhibit was staged (not for competition) by that veteran beekeeper, Mr. SW Marriott, of Nottingham, in the shape of a beautiful grotto built in wax by the bees, in a large feeder. It appears that Mr. Marriott has used the feeder as a receptacle in which to place the cappings of the combs when he has been extracting, then putting the whole over a stock of bees for them to clean out, and carry the honey below to store for winter. They have not only done this, but have built a beautiful grotto in comb work with the wax cappings, which were placed in the feeder and in doing so they have constructed nothing but worker cells. Not a single drone cell could be seen. This instinct in the bee teaches it to carry out an interesting part of the domestic economy of the hive. Drone cells are not now required and will not be until the springtime comes so to avoid waste and utilize every part, no unnecessary work is done and no drone cells built.

Altogether the show was an unqualified success and the bee department was no exception. The management of the Moorgreen Show and NBKA, under whose auspices the show is annually held, are to be heartily commended and it is certain that if more help was afforded the Beekeepers' Association in this way by the stronger societies of agriculture and horticulture it would be to the mutual advantage of both. The agricultural and horticultural societies in the county might with advantage include bees and honey, etc., in their schedules in the future and in connection with the County Association help in spreading a greater knowledge of apiculture, with all its attending benefits and advantages, among our artisans and cottagers in the county

BBJ 12 Sep 1895
NBKA Moorgreen Show. The bee section of the show proved most attractive to

visitors, who surrounded the "Observatory" hives during the afternoon of the 3rd inst. watching intently the busy animated scene within. The honey shown at Moorgreen was of excellent quality and the show was altogether interesting. Mr. P. Scattergood, jun., officiated as judge of honey and made the following awards:

Best Specimen of Bees of any race, with their Queen, in a Unicomb Hive— 1st, A. Warner; 2nd, G. Marshall; 3rd, H. Wiggett

Twelve 1-lb. Jars Extracted Honey — 1st, G. Marshall; 2nd, W. Lee; 3rd, SW Marriott, Nottingham

Best Twelve Bottles of Honey (members with apiaries within five miles of the show ground) — 1st, W. Brooks, Eastwood; 2nd, A. Warner, Moorgreen; 3rd, G. Reeve, Moorgreen

Best Frame of Honey (in case)—1st, Marshall, Norwell; 2nd, SW Marriott, Nottingham

(also reported in Nottingham Daily Express, September 4, 1895)

NEP 18 Sep 1895

The Duke of Devonshire opened the Midland Dairy Institute at Kingston Fields, Notts. yesterday afternoon. A college has been founded by the County Councils of Derbyshire, Leicestershire, Nottinghamshire and the Lindsey Division of Lincolnshire. Some farm buildings on Lord Belper's estate were obtained on highly advantageous terms and these have been adapted at a cost of about £300 to the purposes of this dairy school. Not only is teaching to be given in the manufacture of butter and all sorts of cheese, but the whole management of a dairy farm will be studied, besides poultry and beekeeping.

BBJ 26 Sep 1895

Leicestershire BKA held its first show in connection with the Loughborough Agricultural Association, on the grounds of WB Paget, Esq, Smithfield, Loughborough, on September 18. This innovation was much appreciated by visitors and proved most successful. The manipulating tent was in charge of Mr. AG Pugh of Nottingham, who gave practical demonstrations and lectures on bee-keeping to a very attentive audience. He also officiated as judge.

NG 28 Sep 1895

Last week but one, our neighbours in Derbyshire held the annual County Agricultural Show and, as usual the Derbyshire BKA was represented in the bee department of the show, at which they usually hold the annual show of the association. There were 88 entries in all the classes, and the two for observatory hives were well filled; but taking everything into account, the show was scarcely as good as the Notts. County show held at Newark on Bank Holiday.

Another of our neighbours, viz., the Loughborough Agricultural Association held its annual show last week and the Leicestershire BKA held a very interesting exhibition of bees, hives, honey and appliances. Lectures on beekeeping were also given by Mr. AG Pugh, the secretary of the NBKA. The weather was splendid and there was a large and fashionable company present. The lectures, etc., in the bee tent were well patronised. We would suggest that our Leicestershire friends should do a little more aggressive work in this direction and try to secure the assistance of the County Council to help them in this matter.

A few weeks ago we advised the sowing and planting of several kinds of plants for the benefit of the bee in the spring time. We would now advise the planting of crocus bulbs in fairly large quantities. They supply large quantities of pollen in the early spring when it cannot be obtained from other flowers. A small quantity of nectar is also obtained from crocuses. Another plant which it is desirable to have in the garden also in fairly large numbers is the old fashioned single variety of wallflower. Both of the above not only add brightness and cheerfulness to the garden in the spring time, but provide the honeybee with many a delicious mouthful. They are both comparatively cheap and easy to grow. The writer saw the other day in a public park some distance from Nottingham some thousands of young wallflowers planted for spring decoration.

BBJ 3 Oct 1895
Do Bees Carry Eggs? In reply to Mr. W. Woodley (p. 364) on the above question, I would refer Mr. W. to my letter (p. 345) in which I distinctly state that the queen was received in one of "Mr. Brice's" cages and this was the cage used. I have no doubt that Mr. W. will remember that it was fully described and illustrated on p. 235 of the Journal and if he will kindly read the description and look at the illustration I have no doubt he will clearly see through the arrangement at a glance. At present I am afraid that he is just a "wee bit" mixed in the cage, and confounds this cage with the Root and Howard cages, which I believe some time ago he said, in "Notes by the Way," he uses.

I have always used the 'Howard' cage up to this time and the arrangements are somewhat different, because, in the 'Howard' cage the bees eat away the food from the feed-hole and the queen walks out to her duties, but in the 'Brice' cage after the cardboard slide in the bottom of wooden portion of the cage is withdrawn the queen and the few workers caged with her enter the wire mesh or perforated compartment, which, as I said in my letter, was pressed well into the comb and honey. Consequently there was no ingress or egress via the feed hole into the hive and I expected the bees would eat their way into the cage and thus release the queen, but this they did not do, though they had commenced to do so when I liberated her on the Wednesday.

This, I think, disposes of the second question: Why the queen was caged from Friday till Monday, i.e., just because the bees did not liberate her? Besides which, as I said in my letter, I do not care for too much fussy manipulation. I think Mr. W. has made a slight error here. I said the queen was caged from Friday till Wednesday, not Monday and on p. 346 I gave my reasons for keeping the queen caged so long.

The question of queen re-entering the cage is rather far-fetched and wide of the mark, because she had never been released, consequently she could not go back again into the cage. I am pleased that Mr. W. is trying to get more light on this interesting subject, and, in my opinion, he would see more light if he had for examination one of Mr. Brice's cages.

I hope to meet friend Woodley at the Dairy Show and the quarterly meeting of the British, and then we may have a chat on this and other matters.
Peter Scattergood, Jun., Stapleford, Notts, September 30, 1895.
P.S. — I would have replied before, but have been from home enjoying a much-needed rest at the seaside

Minute Book – Quarterly Committee Meeting held at the Peoples' Hall, Nottingham on October 6th, 1895. Present Messrs P Scattergood (Chairman), H Brooks, JT Faulconbridge, E Forbes, SW Marriott, W Poxon, AJ Turner, J Rawson, A Warner, F Hallam, G Wood and Hon Secretary AG Pugh.

The secretary read the minutes of the last meeting which were confirmed. The quarterly balance sheet shewing a deficiency of 17s 8d was read. Mr Marriott moved and Mr Faulconbridge seconded that the balance sheet as now read be adopted.

Mr Marriott moved and Mr Turner seconded that first prize for wax from Newark be preserved until the annual meeting.

The secretary read a list of members who are in arrears with their subscriptions which shows that 56 had not paid on September 20th and only six new members had joined and he pointed out that if membership was to be kept up, each committee and other members must do their best to get in the arrears and enrol all new members possible before the years end.

The secretary also gave a short report on the shows held during the season which have all passed off in a satisfactory manner and asked for suggestions for making the next annual meeting as attractive and as popular as possible but none were made except that Mr Hallam thought a leaflet might be advantageous for distribution amongst beekeepers who are not members, setting forth the advantages of becoming such.

Mr Scattergood moved and Mr Turner seconded that the secretary be asked to defer the date of his resignation until March 30th, 1896 so that the appointment of a new secretary may be made by members at the annual meeting. Mr Marriott nominated Mr G Hayes, Mr Turner nominated Mr Scattergood. Mr Pugh nominated Mr W Raven as suitable persons from whom to select a new hon. Secretary and meeting terminated about 4.30pm.

BBJ 17 Oct 1895
Meeting of the BBKA Council held at 105, Jermyn-street, SW, on October 10th. Present: Thos. W. Cowan (chairman), and amongst the ex-officio members were P. Scattergood jun. and AG Pugh (hon. sec. NBKA)

BBJ 17 Oct 1895
The Twentieth Annual Show of the Dairy Farmers' Association was held on the 8th inst.
Twelve 1-lb Jars Granulated Honey (23 entries) — hc, H. Merryweather, jun., Southwell, Notts.

BBJ 17 Oct 1895
Can Bees Distinguish Between Drone And Worker Eggs? From recent correspondence, I suppose we can accept it as a fact that "bees do carry eggs." In Mr. Brice's "Jottings" (p. 333) we read: "Drones are smaller, owing to their being raised in worker cells." This suggests the question, have bees power to discern between drone and worker eggs?

If not, I should like to ask your correspondent, have they been known to carry drone eggs into a queen-cell and attempt to raise a queen from it. If bees can discern the difference between drone and worker eggs, I ask, where the combs in brood-nest consist entirely of worker-cells and there are drone-cells only in super, have bees been known to carry eggs through excluder into drone-cells and so defeat our modern hive arrangements? A Worker, Long Eaton, October 7.

NG 19 Oct 1895
Last week the writer visited the dairy show at the Royal Agricultural Hall, Islington. There was a splendid show of honey, and although that staged by the Notts. bee men did not secure a prize, it was V.H.C.

BBJ 24 Oct 1895
Mr. Scattergood exhibited a cheap feeder to members of the BBKA conversazione, which was made by an ingenious bee-keeper in Derbyshire. It was manufactured out of an old treacle-tin and had been used with good results. He also referred to a matter which would probably interest those concerned in the spread of education. Being officially connected with the School Board in his own district, he had asked the teachers to introduce bee-keeping in the form of object lessons in the schools. As a

result the subject was now taught there and an observatory hive had been shown to the children, which was a capital way of securing their interest. He had also procured sets of cards illustrative of bee-life in some form or other, which on the Kindergarten plan were designed to be filled in with needlework or coloured and he exhibited some that had been so manipulated by little children under six years of age. Such object lessons were allowed by the Code and could be taken as part of the ordinary curriculum.

BBJ 24 Oct 1895
The Dairy Show. Extension of Prize List for 1896. I entirely agree with Mr. Brice (p. 416) "that the Dairy Show should not only be the great honey fair of the year, but the annual gathering ground of British bee-keepers" and if as great an improvement as was shown between this year's show and last continues, there need be little fear that this will soon be the case. What is wanted to more fully popularise the honey classes is well set forth in Mr. Till's able letter on page 417 i.e., "a more comprehensive schedule with far more classes and prize money." If, however, heather honey has a distinct class, there should, I think, not be separate classes for Scotch honey.

What friend Brice terms our "plot" will I hope meet with generous support, in which case there should be little difficulty in getting these additional classes included in the schedules for next year and if the ladies take the honey-confectionery into their special care, no doubt it will be well to the front. I consider the honey shown was much superior to that of last year, nearly every sample in the large class of extracted honey being what might well be termed "show honey" - in, fact, I was pleased to recognise some old acquaintances that I had seen as prize-winners at county shows before.

Had your correspondent, "Contempt of Court" (p. 418), been an observant visitor at bee and honey shows, he would not say "colour takes a subsidiary place." We are all agreed that flavour is the most important factor, but if he had gone carefully round the exhibits at any large show he would find that the high class honeys do not vary so much in flavour as they do in colours and that the latter appeals so forcibly to the eye that he may venture to predict that, in nine cases out of ten, the light amber-coloured honey take premier honours. Quite rightly, too, I think, because it has invariably been gathered in virgin combs just in the height of the season and is not an admixture of good, bad, and indifferent products, as is very often the darker coloured honey.

No doubt Hunts. honey has "taken the cake" this year and if it were desirable and practicable to set up a standard for honey par excellence, I fancy some that has been gathered in Hunts. and shown this year might be used for the purpose; for, personally, I never saw honey of greater all-round excellence than some shown by Hunts. bee-men in 1895. A class for heather honey will enable us to compare such samples as Mr. Webster and Mr. Jacomb-Hood refer to in their letters with the real "native" Scotch article.

Hoping Mr. Till will not have to lament of "lack of public spirit among bee-keepers, or too few prizes at future Dairy Shows."

Arthur G Pugh, Beeston, Notts, October 21.

BBJ 24 Oct 1895

Do Bees Carry Eggs? Our Friend Woodley in his "Notes" (p. 404) gives a good illustration of the old Proverb

> A man convinced against his will
> Is of the same opinion still.

I am afraid it is no use turning on the "search light" in this case, because I think when Mr. W. gets an idea, his methodical mind must have clear light, full proof, and certain knowledge, before he will move from the position he has taken up. I admire him for this; if more bee-keepers were like him, fewer miserable mistakes and canards would be recorded, to the annoyance of modern bee-men.

I am pleased that Mr. W. does not accept the position, in a shut your eyes, open your mouth and take what I give you sort of fashion, but wishes for more light. He must, however, pardon me for turning on the search light to the points raised by him as to the release of the queen, and the fertile worker theory. I have distinctly stated all along that the queen was not set free till I myself released her; she could not therefore slip in or out either "of her own free-will." Then as to a fertile worker, I would ask where did the eggs that produced the worker bees come from if they were not carried into the cells by the bees?

To my mind the question is so clear that no search light, however brilliant, could possibly make it clearer, perhaps Mr. W. may have the opportunity next season of testing this interesting fact for himself.

Peter Scatterqood, jun., Stapleford, Notts, October 17, 1895.

NG 26 Oct 1895

The general secretary of the NBKA wrote the other day asking me to call the attention of our readers to the fact that the committee who have charge of the County Council work, are making arrangements for lectures during the coming winter at various centres on bees and bee-keeping, and as the number is necessarily limited owing to the smallness of the grant, those who make early application will have a prior claim. Hence it is desirable that those who wish a lecture in their neighbourhood should apply at once to Mr. AG Pugh, Mona-street, Beeston. The Technical Instruction Committee and the County Council act wisely, we think, in providing these useful and instructive lectures on bee-keeping at various centres and we could wish the grant were larger so as to meet the wishes of those who cannot now be served, either by a winter lecture or a summer demonstration. As the committee of the Council have so many calls in other directions the grant is of necessity limited, but we are quite sure that the Council has full value for the money voted for technical instruction in bee-keeping.

NG 7 Dec 1895
The other day the General Secretary of the NBKA, wrote and asked us to call the attention of our district to the fact that to all members joining now the subscription will be counted as for next year. If any of our readers are thinking of joining the association (and the writer strongly advises them to do so, because of the mutual helpfulness association brings), they should send on their names and subscriptions to Mr. AG Pugh, Mona-street. Beeston and he will enrol them as members.

This note gives us the opportunity of calling attention to the association itself, as it is doing a good work amongst a large number of bee-keepers in the town and county. Nearly all of the leading members work for the good of the cause without any reward whatever, the secretary being the only paid official and his is only a very small remuneration. The secretary says that the association could and would do more work if it had a larger income. Its chief aim is to endeavour to raise the pursuit of beekeeping from a neglected and degraded position to that of an art, deserving the attention and study of intelligent and cultivated minds, as well as to provide cottagers, agricultural and other labouring classes with the knowledge of improvements in bee culture, so as to enable them to add materially to their incomes. But it cannot carry out this programme as extensively as is desired, owing to the limited funds at its disposal.

From the report we gather that the Duke of Portland, Viscount St Vincent. Lord Newark. Mr. JE Ellis MP, Mr. FC Smith, JP, Mr. AH Chambers, Mrs. J. Hind and others subscribe to the association, but the bulk of the subscriptions are given by the members themselves. We would call the attention of the leading ladies and gentlemen of the county to the desirability of helping on so good a work as to encourage the industry of beekeeping amongst the cottagers and labourers and thus add to the income from the land and create a fresh interest in rural life.

BBJ 26 Dec 1895
Preventing Swarming. Referring to the letter of Mr. Simmins (p. 506), I notice his nadir, or lower chamber, is not added until the second super is placed in position and then he says: "Little or no comb building will be carried on there" (in nadir). Now, I ask, how would a hive "conquer" the invasion of their enemies with a $1\frac{1}{2}$-in. entrance? I know of a well-known bee-master who asserts that comb-building is the real value of a nadir. These statements do not fit very well together. I should, therefore, like to ask your readers to say of what practical value is a nadir? Personally, I always thought that the old Scottish system was to add a nadir first, to give room and yet retain all

the heat in brood nest should the weather be cold.

If the weather became hot, the bees have plenty of spare room to work in, either to store honey or build comb, without fear of brood getting chilled, as is too often the case when a super is given too soon. The modern apiarist tries by foundation and extracting to save all comb building he can. Without more knowledge on the subject, it appears to me a wrong system to give bees the pleasure of building what comb they like; if a nadir is given filled with combed frames, will bees store honey there if requiring room, instead of swarming? In other words, is it comb building or room to store honey that will prevent swarming?

"A Worker", Long Eaton, December 10.

NG 28 Dec 1895

County Council Lectures on Beekeeping. If any of our readers residing in a district where there is a reasonable hope that a lecture on "Bees and Beekeeping, or how to manage them for pleasure and profit," would be appreciated and likely to do good, especially in rural districts, we would advise them to make early application for a lecture to Mr Pugh, Mona-street, Beeston because the number to be given is necessarily limited, owing to the grant for this subject being rather small, and those who have not had a lecture in their neighourhood should have a chance before those who have.

"Household tales with other traditional remains", Sidney Oldall Addey, MA (Oxon), 1895

An old woman in Ompton, Nottinghamshire, used to take great notice of bees. She once saw her bees "creeping about in pairs as mourners do" for many days. Shortly after that her daughter died.

NBKA Membership			
1886	109	1891	184
1887	62	1892	195
1888	83	1893	203
1889	121	1894	198
1890	167	1895	?

Minute Book – Quarterly Committee Meeting held at the Peoples' Hall, Nottingham on January 4th, 1896. Present: Messrs SW Marriott (Chairman), E Forbes, GE Puttergill, AJ Turner, J Rawson, A Warner, G Wood, AG Pugh (Hon. Sec.)

The secretary read the minutes of the last meeting which were confirmed.

The secretary then read the quarterly balance sheet and Mr Turner proposed and Mr Wood seconded that the same be accepted.

The secretary read the annual balance sheet which shows a balance of £1 19s 6d. Mr Warner proposed and Mr Wood seconded that the same be adopted, printed and circulated amongst members.

Mr Warner proposed and Mr Turner seconded that the annual meeting be held on February 22nd, 1896. Mr Marriott proposed and Mr Rawson seconded that a meat tea be provided at the same time and prices as in former years. Mr Turner proposed and Mr Faulconbridge seconded that the secretary be empowered to provide provisions required and have crockery, etc. for 7 dozen persons. Mr Wood proposed and Mr Warner seconded that the annual prize drawing take place as usual and that the secretary provide the following prizes:
1st X-all hive
2nd Cottagers honey ripener
3rd Large rapid feeder
4th A smoker
5th 1lb foundation

Mr Puttergill proposed and Mr Wood seconded that the secretary be asked to write and, if possible, arrange for Mr Simmons to give a lecture after tea at the annual meeting, failing him Mr Scattergood be asked to give his promised paper on "Foul Brood".

A conversation then took place on general details of the annual meeting and election of officers and secretary was instructed on same. Mr Marriott kindly promised two articles for prize drawing.

BBJ 6 Feb 1896

RSP (Newark). Honey not granulated. Honey received, though not yet granulated is certainly granulating. No doubt the temperature of room in which it has been kept has helped to retard its becoming solid as soon as it would otherwise have done, but the weather conditions under which nectar is gathered have much to do with the variableness of honey granulation in different seasons.

NG 8 Feb 1896

"A Country Bee" writes:
(1) Can you inform me why there has been no bee department and exhibition of honey at the Notts. County Agricultural Society's annual show for the last few years?

(2) How and who can revive the same?
(3) Is it not a rule for the County Agricultural Societies show and the County Beekeepers' Association's show to be held at the same time and place?

In reply to your first question, we are not acquainted with the working of the County Agricultural Society, but believe the reason why the exhibition of honey, bees and appliances was discontinued was owing to the necessity for economy which was felt by the Agricultural Society's Council some few years ago. The council could not see their way very clear to give the usual amount of £10 towards the prizes given by the County Beekeepers' Association. We think this was a mistake, because in these days, when we hear so much about agricultural depression, it is well for so many economies belonging to and connected with agriculture to be encouraged and stimulated. A few hives of bees will probably yield for the farmer or cottager a greater return for the capital employed than perhaps any other stock on the farm or homestead besides very materially helping the fruit and flower crop.

Second, as to how we can revive the exhibition? So far as we know, only the council of the Agricultural Society. If any members of the society request the council to consider the advisability this year including bee, honey, hives and appliances, etc, in their annual show schedule we have no doubt they will receive due and courteous consideration both from the secretary and the council. We shall be glad to hear the result of any application.

Third, as to the County Agricultural and the County Beekeepers' Associations show being held at the same time and place. Yes, it is in many counties the rule. We have visited a number of county agricultural shows in different parts of the country and have invariably noticed that the County Beekeepers' Association have held their show at the same time and place. And, further, that the larger society, the Agricultural, have helped with the prizes besides providing staging and tents. They in turn reap the admissions fees, etc. We are glad that this question has been raised, and hope the council of the Agricultural Society and the NBKA may ere long recommence this useful and helpful adjunct to the county show and especially if lectures can be given and more of our rural and agricultural populace helped to manage bees for profit.

NG 22 Feb 1896
The principal note this week must be with reference to the annual meeting of the NBKA at the People's Hall, Heathcote-street, Nottingham. The president, Viscount St. Vincent, announced to take the chair. After tea a talk is to be given on "foul brood," to be followed by a discussion. After this will be the usual prize giving. The association are to be congratulated on the balance and statement of accounts for the year. A brief perusal of the same shows that the Association have been doing good work during the year that six shows have been held under the care of the association,

and that the sum of £3 19s 6d has been spent in prizes. At the above shows the total receipts from all sources are £75 7s 7d and the expenditure to £73 8s 1d leaving in the treasurer's hands a balance £1 19s 6d.

Minute Book – Annual meeting held at the Peoples' Hall, Heathcote-street, Nottingham at 3pm on Saturday, February 22nd, 1896

The President, Viscount St. Vincent, took the chair at 3.10pm - a large number of members being present. After a few remarks by the chairman who congratulated the members upon the improved financial position of the association, the secretary read the minutes of the last meeting which were duly confirmed.

The secretary read the annual report which was of a fairly satisfactory character for, although the membership has decreased and the past honey season could not be called a good one, yet the societies financial position has improved and the shows were carried out with great success. Mr Wootton moved and Mr Marriott seconded that the report as read be accepted.

The secretary read the annual balance sheet and compared the same with the preceding year. Mr Scattergood moved and Mr Riley seconded that the balance sheet as now read be adopted and printed with the annual report.

His Lordship suggested that the election of the committee should be 'en bloc'. The secretary gave particulars of attendance. Mr Hayes moved and Mr Marriott seconded the committee for 1895, with the exception of Messrs Elliott and Webster, be re-elected for 1896 namely Messrs Brooks, Faulconbridge, Forbes, Marriott, Merryweather, Poxon, Puttergill, Turner, White and the hon. secretary and experts.

Mr G Wood moved and Mr Pugh seconded that Mr Mckinnon of Epperstone fill the vacancy on the committee caused by Mr Elliott's resignation. Mr Marriott moved and Mr Cartledge seconded that Mr Annable of Eastwood fill the vacancy on the committee caused by Mr Webster's inability to attend.

Mr Glew moved and Mr Herrod seconded that the present sub-committee for technical education in beekeeping, ie. Messrs Pugh, Raven, Marriott, Riley and Scattergood be re-selected.

Mr McKinnon moved and Mr Warner seconded that Mr AG Pugh be asked to kindly act as representative of NBKA at the Council meetings of BBKA on the same terms as before. Mr Wright moved and Mr Wood seconded that the general secretary act as their representative when required.

BEEKEEPING IN VICTORIAN NOTTINGHAMSHIRE

Mr Pugh moved and Mr Turner seconded that this meeting most heartily thank his Lordship, Viscount St. Vincent, for his past services as president of this association and cordially invite him to accept this position in 1896. (Carried with applause). His Lordship having very kindly accepted the position, assuring all present that he had great sympathy with the work of the association and hoped to continue his connection with NBKA for many years, the election of hon. secretary was proceeded with. Mr Raven having withdrawn his nomination there was only Messrs Hayes and Scattergood for election and it was decided this was to be by ballot. The candidates retired and Messrs Mills and Herrod, having been appointed scrutineers, announced that the voting was as follows: 24 for Mr Hayes and 15 for Mr Scattergood, two papers having been spoilt. The President declared that Mr Hayes to have been duly elected.

Mr Marriott proposed and Mr Turner seconded that Mr AG Pugh be elected hon treasurer.

Mr Riley proposed and Mr Herrod seconded that Mr Scattergood be thanked for past services and be hereby re-elected as auditor for 1896.

Mr Scattergood proposed and Mr Wood seconded that the committee shall, in future, be termed 'council'. This was met by an amendment that it remains as at present.

Mr Wood proposed and Mr McKinnon seconded that receipt books to be provided for district secretaries and experts and that they may give an official receipt for all subscriptions received by them.

At about 4.50pm nearly 80 members and friends sat down to tea together and at 6.20pm the President again took the chair and invited discussion on any subject members thought it desirable to ventilate; some remarks were made about different sized bottles being exhibited in honey classes at the annual and other shows, the general feeling appeared in favour of adopting a standard size, which by resolution was passed.

It was decided that an open class for both run honey and sections should appear on the schedule of the next annual show, also that novice classes should again be adopted and that NBKA pendants should be offered in prizes and that anyone who had already won one should have the option of taking the cash value in lieu of same. The committee were recommended to renew the rule relating to first prizewinners being debarred from showing in similar classes (honey) at other local shows during the season. It was also decided to keep the first prize wax till the next annual meeting.

Viscount St. Vincent having to leave early to catch a train, Mr Wootton was elected chairman. Mr Scattergood proposed and Mr Meadows seconded that a vote of thanks to Viscount St. Vincent for his services as chairman. After it had been responded to, Mr Scattergood read his paper on 'Foul Brood' which was listened to with much interest by all present. A long discussion took place and the time having so far advanced, it was decided to defer reading the paper which Mr Simmons had sent on his non-swarming system till a future time. A vote of thanks was, however, given to Mr Simmons for his kindness in preparing and sending his paper.

The meeting closed with the usual prize drawing:

1st Prize	Wells' hive	given by President	won by Mr Warner
2nd	'XL All' hive	given by President	won by Mr Meldrum
3rd	'XL All' hive	given by NBKA	won by Mr Mills
4th	Cottagers honey ripener	given by NBKA	won by Mr Measures
5th	Rapid feeder	given by NBKA	won by Mr Jebb
6th	'Beeston' Feeder	given by Mr Puttergill	won by Mr Clark
7th	Rapid feeder	given by Mr Marriott	won by Mr Gray
8th	Bee escape	given by Mr Marriott	won by Mr Temple
9th	100 Sections	given by Mr Pugh	won by Mr Marshall

NG 29 Feb 1896

"Foul Brood" paper read at the annual meeting of NBKA on Saturday last:

Ladies and Gentlemen: My sole aim and object in agreeing to introduce this subject at this time is to make a greater interest in this, which to beekeepers is a very important question, and if possible to try to back up and strengthen the hands of the BBKA in their laudable effort to obtain legal powers and enactments to work towards the stamping out of the disease, by legally dealing with carelessness and stupidity and ignorance. During the last few months no subject has been brought so prominently before the beekeeping fraternity and the public at large, through the press of all shades of opinion, as this question of Foul Brood amongst bees. So, if apology for bringing up this question is needed from me, it is that the awakening of the public interest and the feeling of our beekeeping industry, and the helping of our poorer brethren of the craft or cult.

There are some people today who lay all the blame of this pest on our modern methods of beekeeping but I scarcely think this is fair. Years ago bees were rarely taken more than a mile or two from their native spot and stocks "taken up" in the autumn were consigned to the sulphur pit (which in itself must be a capital disinfectant) and, in the majority of cases, afterwards the straw skep consigned to the garden bonfire. If foul brood lurked here and there, as I am inclined to think it did, to a large extent by these methods and means, it was perhaps unconsciously kept in a narrow area and localized and the facilities for its propagation and spread, to a certain extent, were limited; but the conditions are very different today. Bees are sent through the length and breadth of the land and sometimes sent into a healthy district from an infected one with the disease and thus it is passed on till a clean area is rare and beekeepers are fortunate residing in those areas where foul brood is not.

The question arises naturally here. What is foul brood? It is a kind of minute vegetable organism commonly spoken of as a germ. It has been scientifically described: "Bacillus of foul brood occurs in the tissues and juices of bees and especially their larvae". These bacilli are rod-shaped and if you suppose a round ruler to elongate without increasing in thickness and then at a definite point break in two and then again in like manner, you will have a fair idea of how the bacilli multiply. Once the bacilli have exhausted all the nutriment around them they turn to spores or seeds and whilst it is possible to check the disease in the bacilli stage, it is almost impossible to do so in the spore stage without killing the bees.

But even this description may not be plain and simple enough for some and they may say, "This may be alright; but tell us how it appears, how does it get in the hive, and how may we distinguish it, how we may know it when we see it?"

To such, and indeed all, I would say, "Purchase the pamphlet 'Recognition and description of Foul Brood' from the BBJ Offices costing one shilling and you will have at a glance and always to hand a clear, concise and accurate description of the disease; but for benefit of those who will not purchase the aforesaid monograph, let me give you a description, largely taken from the book quoted above.

If a comb is removed from a healthy brood nest during the breeding season, the brood from such a hive will be easily recognised by being compact and the larvae in its various stages will be of pearly whiteness and they will lie curled up in the base of the cells, something after the form of a letter C whereas in a hive affected by bacillus where the larvae affected by the disease move uneasily in their cells and instead of being curled up and plump, they lie lengthwise in the cells and be flabby in appearance. They will change in colour to yellow or faint buff, changing later to brown and begin to decompose; later they dry up, leading to the body being shrunken by evaporation, leaving a dark, almost black mass on the lower side of the cell. Should the larvae escape the attack of this disease till near the period of pupation, it is sealed over in the usual way but the inhabitant of the cell is bound for death and so here and there cells will be found with cappings rather dented and darker than healthy brood and punctured or perforated with holes of irregular sizes. If the cappings of the cells are removed and an attempt to remove a dead larva from a cell where is has assumed the coffee coloured state, a putrid, ropy sticky mass will be found to be the remains of the larvae and this often give off a foul odour, which in bad cases may be smelt several yards from the hive. If it gets this bad then a cure is almost impossible.

Assuming that my description is now plain enough for all, I would in the next phase call your attention to some of the methods under which the disease spreads and is propagated. These are very varied and, perhaps on the whole and in detail, are not fully understood.

Disease is generally caused by weakness and any tendency to weakness, either through chilled brood, weak stocks or in-and-in breeding, should be guarded against. Weak hives have no chance against the scourge of the disease, while old mouldy combs, damp hives, etc. are a most active medium for the propagation of it. When stocks become weak through the disease, other colonies rob it and these spread the disease by carrying the germs home with their plunder. The grubs are probably most possibly affected by the antennae of the nurses, who are continually travelling in the darkness of the hive, near the comb and inserting their antennae into the brood cells and thus become aware of the condition and needs of each occupant and in doing so the antennae must continually (where disease reigns) be coming into contact with larvae full of bacilli and also into contact with the foul, sticky masses into which the larvae change a few days after death. Probably while thus being brought into contact with the bacilli, they may remove spores (or seeds) and this, transferred to the next grub will find suitable material in which to grow and develop and thus spread the disease.

The beekeeper himself may also become a source of infection. He may, and in many cases no doubt will, carry on his hands, clothing and appliances used for manipulating a diseased stock, the spores or bacilli and even hours afterwards transfer them to a healthy stock and thus spread the disease.

BBJ 5 Mar 1896
The annual general meeting of the NBKA was held at the People's Hall, Nottingham, on February 22, 1896. Viscount St. Vincent (president) in the chair and among the good attendance were Messrs. G. Hayes, P. Scattergood, Herrod, Glew, Wootton, Marriott, Wood, McKinnon, Richmond, Raven, Forbes, Baguley, Rawson, Riley, Meadows, Puttergill, Newton, Rev. JS Wright, etc.

The Chairman, in briefly opening the proceedings, remarked that that was the first time since he had been chairman at their annual meeting that the association had a balance on the right side.

The minutes of the last meeting having been read and adopted, the Hon. Secretary (Mr. AG Pugh) read the annual report, which, after remarking as to the season of 1895, and dealing with membership and income for the past year, went on to say:
It was proposed to give lectures on bee-keeping during next month at East Leake, Collingham, East Bridgford, Underwood, etc. That scourge of bee-keepers, 'foul

brood,' had received considerable attention at the hands of the Council of the BBKA, and it was reported that proposed legislation dealing with the subject was making good progress. Their experts visited ninety-nine members' apiaries during the season, and did not report any great increase of the pest in Notts.

The report, along with balance-sheet, which showed a balance in the hands of the treasurer of £1 19s 6d was adopted.

On the proposition of Mr. Pugh, Viscount St. Vincent was re-elected president. Sergeant McKinnon was elected to the committee. The sub-committee on technical education was re-elected. Mr. Geo. Hayes (Beeston) was elected secretary by ballot, Mr. Pugh having resigned his dual post of hon. secretary and treasurer; he, however, agreed to meet members' special wishes, and accepted the position of hon. treasurer.

Tea was afterwards partaken of, after which a conversazione took place, at which Mr. P. Scattergood read an instructive paper upon "Foul Brood." A discussion followed, and the proceedings were brought to a close by the usual prize drawing (for which a valuable collection of appliances had been provided), included a splendid "Wells 7 ' hive and an "XL all" hive presented by the noble President, Viscount St. Vincent.

George Hayes was secretary of NBKA for many years. He wrote the authorative book, "Nectar Producing Plants and their Pollen" a copy of which he donated to the NBKA library. He contributed many articles to BBJ. He donated all his beekeeping books to NBKA and their library was named after him. It is still used by the members today including the author in preparing this book.

NG 7 Mar 1896
Foul Brood (Continued). The disease being of such an infectious character, and the beekeeper himself being such an unconscious cause of spreading it, I can well understand some of you saying, "What about experts? Are they not very likely to carry the disease from one apiary to another?" This is indeed a serious question and I am almost inclined to think that sometimes that it would be safest, if in some districts there was less expert work done (ie. so far as actual manipulations are concerned), because by this means the disease may be carried from one apiary to another, even though they may be several miles apart.

An expert, when he discovers the disease in an apiary, should instantly acquaint the beekeeper and gave him the best advice he possibly can and help him to adopt some of the methods suited to the case to, if possible, stamp out the disease. After his examination and manipulation (and this applies to all beekeepers as well as experts), he should thoroughly cleanse his hands with strong carbolic soap, or in a weak solution of mercuric chloride, and spray his smoker and veil with a solution of phenol It would also be a wise precaution if he did not manipulate or examine anymore hives till his clothing and appliances have been thoroughly fumigated. If an expert has the disease in his own apiary, he should not be allowed to visit and manipulate at other people's apiaries and if be has not the good sense to know, he should be told and very plainly by those in authority.

I have been asked what is the difference between chilled brood and foul brood, and how may I distinguish between the two? These are very natural questions and I will endeavour to answer them. It requires come care and discrimination to distinguish between chilled brood and foul brood, because some of the symptoms may appear in the absence of the disease when the brood has died from chill. The larvae in chilled brood, however, is usually grey and not brown as in foul brood, the stringy or "ropy" character in foul brood is absent in chilled brood. Sometimes, however, nothing but a microscopic examination will settle the point beyond the possibility of a doubt and if a carefully made package containing a piece of comb with brood in several stages is sent to the BBJ Office, with a respectful request for information and the usual postage for reply, the editors will send a reply, or publish an answer in the "BBJ" as to whether it is foul or only chilled brood.

Supposing a beekeeper has unfortunately discovered that he has this pest in his apiary, what is the best and cheapest course for him to adopt? That will largely depend upon the circumstances of the case. Let me say here (and forgive me if I appear to be dogmatic) whatever is done must be done thoroughly and well. There must be no tinkering, no half-hearted work, no fitful spasmodic or slipshod action, but the thing must be dealt with sharply and followed up if we are to be successful Unfortunately, however, great ignorance prevails on the subject and not one beekeeper in twenty can tell foul brood when he sees it in its earliest stages and when it is easiest to be dealt with. I am convinced many beekeepers have the disease lurking in their apiaries without their knowing it. Especially is this so where the bees are kept in straw skeps, because of the difficulty of careful and thorough examination. The straw hive may become a hotbed for the disease and owing to its construction, presents the worst possible features and conditions for dealing with it; again, how often do we come across in some sort of the way, place men who call themselves beekeepers, whose hives are all straw skeps and these in such a dirty, slovenly, untidy condition, courting disease and a standing danger to all beekeepers round about.

Now as to remedies. Many and varied are the supposed cures and remedies advertised, but in my judgement the cheapest cure, if the disease has obtained a firm footing and the bees are only a few in number, is to smother the bees with sulphur and take the frames, combs, quilts, bees, etc., and put the whole of them into a furnace if there is one handy, or make a big bonfire in the garden in the evening and burn the lot; if a straw hive, the hive as well. If the hive is a wooden one and worth preserving, thoroughly scour it inside and out with strong soap and boiling water, in which plenty of soda has been dissolved and when dry, paint thoroughly inside and out with a solution of Calvert's No. 5 carbolic acid and water in equal parts. If the disease is only slight and there are plenty of bees, the best thing to do perhaps is to make an artificial swarm of the bees by placing them in an empty skep and feed them with syrup, which has been medicated with napthol beta (three grains of napthol beta to each pound of sugar used being the usual quantity advised and recommended on the packets). Then take all the frames, combs, quilts, etc. and burn them and proceed to thoroughly clean and disinfect the hive as before advised and after the strong fumes of the carbolic acid have disappeared, it may again be used. Place from four to six new frames with full sheets of good foundation in a clean hive and provide new quilts and coverings and then in, say, from two to three days, put the bees into the hive, continue to feed them with medicated syrup, as before advised (be sure and destroy by fire at once the old skep in which the artificial swarm was hived). If the queen is removed and a young vigorous one given from a clean, healthy stock, it will cause the bees to work with renewed energy and vigour and in a few days it will be possible to discover the condition of the brood. In addition to the above place two balls of napthaline split in half at the end of the hive farthest away from the entrance and renew them from time to time, as they evaporate.

All the manipulations I have mentioned should take place in the evening, when there are no bees flying. I know there are numbers of other remedies advised and suggested; amongst them are camphor, phenol, formic acid, salicilic acid, eucalyptol, thymol, and others, but some of them require more care in administration than the ordinary beekeeper can give and as napthol beta and naphtaline may be obtained in handy, cheap, and convenient packages, with full, dear, and concise instructions and directions with each packet, from the BBJ Office, and all are chemically correct in quantity, I have no hesitation in saying that they are the simplest and best remedies the average beekeeper can adopt in fighting the disease.

But we may talk for an age, we may try to keep our own bees clear of the pest and, perhaps, in the neighbourhood there may be a stupid, cantankerous, cross-grained, old-fashioned drone of a bee man, whose mouldy, dirty old skeps, plastered all over with all kinds of strange compositions, are reeking with foul brood, but who obstinately refuses to accept help or advice and who, with a knowing shake of the head, tells us he knows all about it. Such cases as these make it almost impossible for others to succeed. Our main difficulty is at present that we cannot do anything with such people. I say at present advisedly, because, as I said at the commencement, the BBKA and the representatives of the County Councils are endeavouring to obtain legal protection for the careful beekeeper, by, if possible, bringing into force an Act of Parliament to be known as the Foul Brood Diseases of Bees Act, 1896.

Let it be our object to strengthen their hands, to support them in all our power, and if this is done all over the country very soon we shall obtain for our industry that legal protection against ignorance, stupidity and carelessness which as Englishmen is our right and privilege.

Beta napthol is a slowly volatile crystalline compound which was used commonly to impregnate corrugated paper banbds for placing around the tunks of mature apple trees to kill codling moth larvae. It would be readily to beekeepers.

There would have been several ways of administering Calverts No. 5 carbolic acid but the most common was in a solution of glycerine and water (shaken vigourously to ensure a mixture). A few drops of the solution were then sprinkled on a cloth that was bigger than the top of the brood box and which was placed on top of it. The Bee-keepers' Record of March 1901 advised strongly against using this.

Napthaline (Napthalene) is used to make moth balls.

Leicestershire Chronicle 14 Mar 1896
On Saturday afternoon the 14th annual meeting of Leicestershire BKA was held at the Victoria Coffee-house. Sir Israel Hart presided and amongst those present were EN Lewis, Scattergood (Stapleford), AG Pugh (Beeston), GW Marriott (Sneinton Dale). The Chairman remarked bee-keeping was specially an industry for the young, as the way in which the bees did their work was a lesson of great value. Ald. Underwood seconded this, also expatiating on the lessons that the young might learn from the habits of the bees. Mr. Scattergood, in supporting, advocated the inculcating of bee-keeping in the minds of the young, remarking that it conduces to tidiness and industry.

Minute Book – Quarterly Committee Meeting held at the Peoples' Hall, Nottingham on April 4th, 1896. Present Messrs Scattergood (Chairman), Pugh, Marriott, White, Warner, Brooks, Gray, Rawson, Hallam, Raven, Poxon, Faulconbridge, Belton, McKinnon and Annable.

The minutes of the annual meeting were read. The minutes of the last quarterly meeting were read and confirmed. The quarterly balance sheet was next gone through. Mr McKinnon proposed and Mr Raven seconded that the same be adopted.

The question of localising the '[Beekeepers's] Record' was discussed but deferred to a future time.

Mr Raven proposed and Mr Rawson seconded that a recommendation be sent to the next annual meeting to make Mr AG Pugh a life member in consideration of his service as secretary of the NBKA.

The applications from Hucknall, Southwell and Moorgreen societies for honey shows were considered and accepted. Mr Marriott proposed and Mr Annable seconded that the secretary be empowered to make the best arrangements he can for honey shows and, in particular, bring to a special committee meeting when completed.

A discussion took place as to expert's fees but no alteration was made although the meeting was of the opinion that all visits should be made before June 30th.

It was also arranged that the secretary should clear up with the treasurer at each month end.

Mr Pugh proposed and Mr Rawson seconded that the secretary write to Mr Clark of Retford to ascertain if we can get a show thereabouts.

The question of a delegate to BBKA meetings was discussed but the question deferred to the annual meeting.

NG 11 Apr 1896
SH. wishes to know what kind of hive a "Stewarton" hive is. He has heard of a hive of that name from a Scotch friend, but has not seen one. Are they peculiar to Scotland? The Stewarton hive is co called from the district of Stewartown, Ayrshire, N(orth) B(ritain), where it is said to have been invented about 1819, by Robert Kerr. It is still largely used in Scotland, though not exclusively. We have seen several in Notts, during the last few weeks. It is octagonal in shape, and consists of a series of octagonal boxes, usually four in number, placed one above the other, similar to the tiering up of bar frame hive. These boxes are $13^{3}/_{4}$ in. from side to side, and about $5^{3}/_{4}$ in. deep. The hives are usually made in Scotland by specially prepared machinery, though we believe that several firms of appliance makers in England obtain them for customers. It will prove rather a formidable job for the beekeeper to make one himself, owing to the arrangement of the slides, etc. Some beekeepers claim that the octagonal hives possess an advantage over square ones, in the fact of their being a much nearer approach to the circular form, which they also claim is the one in which the interior warmth is most uniformly diffused, with a freedom from cold corners. Perhaps this is more imaginary than real. We do not, however, advise SH to purchase one, as in our opinion they are only one remove from a skep hive, as the variation in the form of the top bar or slides which the shape of the hive entails, certainly prevents their complete interchangeability and to some extent stereotypes the position of the combs in the hive.

BBJ 16 Apr 1896

The monthly meeting of the BBKA Council was held at King William-street, Strand, on April 10, Mr. TW Cowan (in the chair). In presenting the Education Committee's report, Mr. Cowan stated that arrangements were in progress for the holding of examinations for third-class certificates for Group 1 (Notts, Derby, and Leicester) at Leicester, on June 23.

BBJ 3 May 1896
On Sunday, May 3, a fine swarm issued from a hive in the Park, Nottingham. The owner, a leading gentleman in the town, rode over on his bicycle to fetch me to hive the bees for him, but, being from home at the time, my wife gave him instructions how to proceed and these were carried out by the wife of the gentleman referred to, who bravely hived the swarm into a straw skep from an awkward position. I went over on the following day and put the swarm into a frame-hive and it is now doing very well indeed and working away as only swarms do work. This is the earliest swarm I have heard of in this district this year. P. Scattergood, Jun., Stapleford, Nottingham

BBJ 6 May 1896.
The NBKA held their annual show at Moorgreen in conjunction with that of the Agricultural and Horticultural Society on the 8th inst. when over 900 lbs of excellent honey was staged and proved a very attractive department of the show, the tent being crowded most of the day. Mr. TW Jones, of Etwall, Derby, was the judge, assisted by Mr. SW Marriott, of Nottingham, and made the following awards:
Best Hive - 1st, GE. Puttergill, Beeston
Best Hive {made by Amateur) - 1st, JT Faulconbridge, Bulwell; 2nd, JF Simpson, Underwood
Single 1-lb. Jar Extracted Honey (Open) - 3rd, T. Blake, Broughton; 4th, G. Marshall, Notts.
Honey Trophy - 1st, J and W Herrod, Newark; 2nd, G. Marshall, Norwell; 3rd, JWS Rawson, Selston; 4th, JT Faulconbridge
Twelve 1-lb. Jars Extracted Honey (Local) - 1st, JWS Rawson; 2nd, Wm. Brooks, Eastwood; 3rd, GM Bolton; hc, J. Rawson, sen.
Twelve 1-lb. Jars Extracted Honey (Notts only)—1st, G. Marshall; 2nd, J and W Herrod; 3rd, P. Scattergood; 4th, H. Merryweather; 5th, GE Puttergill; hc, JF Simpson, CM Lindley
Six 1-lb. Sections - 1st, G Marshall; 2nd, J and W Herrod; 3rd, GE Puttergill; hc JWS Rawson
Six 1-lb. Jars Granulated Honey - 1st, JWS Rawson; 2nd, H. Wiggett, J and W Herrod
Shallow Frame of Comb Honey - 1st, T. Marshall; 2nd, G Marshall; 3rd, J and W Herrod
Six 1-lb. Jars Extracted Honey - 1st, GM Bolton; 2nd, Geo. Hayes; 3rd, Wm. Poxon; 4th, J. McKinnon
Six 1-lb. Sections (Beginners only) - F. Wygett, Annesley; 2nd, G. Reeve, Moorgreen

BEEKEEPING IN VICTORIAN NOTTINGHAMSHIRE

Twelve 1-lb. Jars Extracted Honey (Blow's Jars) - 1st, GE Puttergill
Twelve 1-lb. Sections (in Blow's Sections) - 1st, JWS Rawson
Honey Vinegar - 1st, GE Puttergill; 2nd, P. Scattergood
Honey Cake - 1st, P. Scattergood; 2nd, J and W. Herrod
Observatory Hive - 1st, H. Wigget; 2nd, J. Annable; 3rd, GE Puttergill; 4th, P. Scattergood
Wax - 1st, G. Marshall; 2nd, GE Puttergill; 3rd, J. Gray, Long Eaton

The proudest day of our lives as exhibitors, noted the Herrod brothers, was when we beat the then Lord St. Vincent in the class for the best display of honey and bee produce, at this show. As can be seen from the results above his Lordship did not take part in the show on this occasion.

NG 9 May 1896
The old adage about swarms may not be strictly true that a swarm in May is worth a load of hay, still a swarm coming off early in May maybe turned to good account and even from the swarm a surplus of honey may be obtained. We heard the other day of a fine swarm, which issued on Sunday May 3rd from a hive belonging to a gentleman in Nottingham Park. It was duly hived by the lady of the house and, on examination of the parent stock, it was found not only to contain a very large amount of brood but also a large amount of sealed honey in the back frames. We think this is one of the earliest swarms in the district this year.

Number of Brood Frames in Hives. A Tomkin asks how many brood frames we would advise in his hive, because he has now six in, and they are full of bees.
In a rich and good locality, ten Association standard frames; in a poor one eight considered about the thing, but the precise number must be left to the judgement of the beekeeper. A stock should only have as many frames as it can cover. If you find that after a while your bees fill all the frames then when the limit is reached, give the supers. Prosperity cannot be judged alone by the number of brood frames.

Minute Book – Special Committee Meeting held at the Peoples' Hall, Heathcote-street, Nottingham on May 16th, 1896. Present: Messrs Scattergood (Chairman), Annable, Forbes, Faulconbridge, Gray, Marriott, Newton, Puttergill, White and Wood.

The minutes of the previous meeting were read and confirmed.

A letter was read from the president stating his inability to attend and offering the same amount for prizes as last year. Letters of apology were read from other members also.

The secretary reminded the meeting that the committee had already arranged for shows with the Lowdham, Hucknall and Moorgreen societies. It was resolved that the terms for the Southwell show be accepted ie. we give 25% of the prize money awarded over and above the £2 given by our president (Lord St. Vincent) to the Southwell society. It was resolved that the terms for the Norwell show be accepted.

It was resolved that the offer of the PSA Newark be accepted and that the county show be held there on August 3rd.

It was resolved that Mr Jones of Etwall, Derbys. be asked to judge at the annual show or failing him Mr J Howard, then Mr Webster. It was also resolved that Mr Marriott be assistant judge at the Newark show and that he be paid 7s 6d for his services in that capacity. It was agreed that judges at the local shows be paid 5s on train fares for each show.

It was resolved that the secretary send a request to the County Council committee for the bee tent to be sent to Newark, Norwell and Southwell and ask them to appoint a judge for two shows. It was agreed that Mr Scattergood be the judge at two shows and Mr Pugh at one and this they are to arrange between themselves which they shall be.

The conditions of the schedule were then gone into and it was ordered to add to No. 3 -"and bottles used must not exceed 6 inches in height" and to No. 5 add - "without consent of stewards". That the bee driving competition be abandoned and a class for sections be added for those members who have not previously taken a prize for sections. Several other minor alterations were pointed out and the secretary requested to make them for new schedules.

BBJ 21 May 1896 Joiners Wanted. Used to bee appliance making.
Apply Walton, Muskham.

Minute Book – Special Committee Meeting held at the Peoples' Hall, Nottingham on Saturday June 6th, 1896. Present Messrs Pugh, Marriott, Measures, Brooks, Warner, Wood, Gray Newton, Faulconbridge, Turner, Mackender, Herrod, Puttergill, Scattergood, Forbes and by special request Mr Mills of Newark. Mr Pugh having been voted to the chair, the secretary read the minutes of the previous meeting. Mr Wood proposed and Mr Newton seconded that the same be adopted.

The secretary then read a letter from his Lordship and from several other members, stating their inability to attend. He also read correspondence from Newark setting forth that the arrangements made for Newark PSA had practically broken down as the PSA people had curtailed their grant. Mr Scattergood moved and Mr Marriott seconded that this association does not hold a show with the PSA people at Newark this year.

BEEKEEPING IN VICTORIAN NOTTINGHAMSHIRE

Mr Scattergood moved and Mr Turner seconded that we try to get to hold our annual show with the Moorgreen show in September and that Messrs Warner and Brooks be appointed as deputations to wait on the Moorgreen people to make the best arrangements they can for this association. Mr Herrod moved and Mr Mackender seconded that, failing Moorgreen, we try the Ossington show at Carlton-on-Trent.

Mr Herrod moved and Mr Puttergill seconded that the open classes for extracted and section honey be taken out and one new class put in in lieu of same for a 1 lb bottle of honey, the honey to be the property of the association as an entry fee and that the prizes in the other classes be 10/-, 7/6d, 5/- and 2/6d.

Mr Scatterhood moved and Mr Warner seconded that in the class for appliances a condition be made that, unless there are three competitors, the first prize will be withheld. Mr Scattergood moved and Mr Newton seconded that the remainder of the schedule remain as at present.

As the grant from the society at ..... was inadequate, several promised subscriptions and the secretary collected from the meeting the sum of £2 11s 6d.

A vote of thanks was passed to Mr Mills for his endeavours to arrange a show at Newark.

BBJ 11 Jun 1896
Wanted. Three-pint SELTZOGENE and Cash for Strong swarm. William Herrod, Trentside Apiary, Sutton-on-Trent, near Newark.

BBJ 11 Jun 1896
On the 2nd inst. I was fetched from business to take a swarm of bees which had settled on some rails in the main part of London-road, Nottingham. When I got there a policeman had been trying to burn the poor bees and to destroy them. He had killed most of the swarm, but fortunately the queen, with some of the bees, escaped and settled nearly at the same place. The people were quite astonished at seeing them taken.
A Constant Reader

NG 13 Jun 1896
APJ calls our attention to the fact that there is no bee demonstration or show of honey or bee appliances at the Notts. Agricultural Show at Colwick. He would like to know why this is so and this is a great mistake. Along with APJ we regret that there were neither bees nor honey at the show but as to why this was so we are not able to say. Especially in these days, when everyone is doing his level best to help the farmer and to obtain as much as possible from the land, something might be done in this way. We are decidedly of the opinion that a major activity like beekeeping ought to be more widely undertaken and helped. The County Agricultural Society and the

Beekeepers' Association, in their capacity, might arrange for this interesting subject to be brought to the front at so prominent a show as that held last week. We sincerely hope that another year some arrangement may be made, so that beekeeping which, without doubt, is a distinct part of agriculture, may be recognised at our county show, as it is at the majority of county agricultural shows throughout the country.

NG 20 Jun 1896

The "Weed Foundation" is made by a generally new kind of machinery. The supply to date gives the best results in the hive of any foundation I have ever tried. It is sold in England at present by only a few appliance dealers at about the same price as our ordinary English make. One English maker of our acquaintance is importing a new machine from America in order to make the foundation at home and with the present system he now makes up several tons of wax every year into foundation. We would certainly advise our readers to secure a supply of the "Weed" with their next order.

I have been frequently asked why does the foundation fall or break down. There are several reasons which may be given. One is, perhaps, it has not been properly fixed in the frame. Another is that the wax by which the foundation is made is of very inferior quality and as soon as the sheet is placed in the hive and the bees start clustering upon it, this raises the temperature in the hive, the inferior qualities will not withstand the extra pressure and so it breaks down. This may be avoided by using nothing but first grade foundation and by properly wiring the frames and securing foundation to the wires.

Our attention has been called to another danger in using poor foundation. A prominent bee-keeper has been out on expert work for the NBKA and he has been shown a sample of comb foundation purchased this year which was nearly black and of very poor quality while the price paid was higher than first-class appliance dealers charge for high grade quality. After it had been placed in the hive and the bees had worked it out, the hive developed unmistakable traces of foul brood, while before the foundation had been placed in no foul brood had been seen or known. This our reader will readily see is a serious question and one in regard to which our appliance dealers should be very aware because in this way disease may be spread far and wide.

The season for showing is fast approaching. The secretary of the County Beekeepers Association informs us that shows are arranged for Lowdham, Hucknall, Southwell, and Norwell. The annual show is to be held in September in connection with the Moorgreen Agricultural Society's Show, while at many of these shows and at others also, arrangements are being made for Lectures and Demonstrations on technical instruction in beekeeping to be given on behalf of the County Council Technical Institute Committee. It is to be hoped that they may prove helpful to all.

BEEKEEPING IN VICTORIAN NOTTINGHAMSHIRE

BBJ 25 Jun 1896
September 8, At Moorgreen. NBKA in connection with the Greasley, Selston, and Eastwood Agricultural Society. Open class for a 1 lb bottle of honey. Schedules ready. Apply Geo. Hayes, Mona-street, Beeston. Entries close August 29.

Minute Book – Quarterly Committee Meeting held at the Peoples' Hall, Heathcote-street, Nottingham on Saturday July 4th, 1896. Present: Messrs Annable, Brooks, Gosling, Gray, Forbes, Herrod, Newton, Pugh, Poxon, Scattergood, Turner, Warner and hon. Secretary. Mr Scattergood was voted to the chair.

The minutes of the previous meeting having been read and confirmed, Mr Pugh proposed that the quarterly balance sheet as read by the secretary be adopted. Mr Herrod seconded and it was carried unanimously.

No other business was done as there had been two special meetings during the quarter.

NG 11 Jul 1896
Through the courtesy of the general secretary of NBKA (Mr. G. Hayes, of Beeston) we are able to give our readers a full list of the honey shows, etc. to be held at various centres in connection with the agricultural and horticultural societies in their respective districts. The list is as follows:
Hucknall Torkard. July 21st; Lowdham, July 23rd; Southwell. July 23rd; Norwell, August ?? and the annual show at Moorgreen on September 8th.

At every show there is a good list of prizes and members of the NBKA are allowed to exhibit at a reduced entry fee.
In addition to the above, the County Council bee tent will be in attendance at Southwell, Norwell, Clarborough and Ossington Show at Carlton-on-Trent, with lectures and demonstrations and manipulations of live bees, showing the most advanced and best methods of modem apiculture, will be given by able bee masters.

BBJ 16 Jul 1896
W. Dollman (Doleman?) (Nottingham). A Novices Queries.
The appearances detailed point rather to loss of the queen through some accident while examining the combs in searching for her. Novices should not overhaul combs so often as once a week. The books named contain all the information we can give you on the points raised, so that when others recommend different plans it is for you to say who shall be the guide.

NG 25 Jul 1896
Honey shows are now taking place, those at Hucknall and Southwell having been

held this week. If any of our readers contemplate exhibiting at any of the remaining shows under the auspices of the NBKA and wish to place a good exhibit on the show table it will be a good plan to select the lightest coloured honey they may have and not to extract from any but sealed combs, strain through flannel to get rid of bits of pollen and bits of comb, then bottle in jars, screw cap or tie over, screw top for preference. Remove all the froth from the top of the bottles and see that all bottles are well placed and wiped when they are staged. Should exhibitors not be successful in their first attempt they should not be disheartened, but try again, and pay great attention to the exhibits that do win prizes, and make a few notes as to the source of the points gained over their own by a successful exhibit. Whatever else exhibitors do not lose their temper because the judge does not award them a prize, and do not let them attempt to set their opinion against the judges, because, as a rule, the latter are experienced men and have no personal end to serve in adjudication, besides which, they have no means of knowing whose honey they may be judging.

NG 8 Aug 1896
This week the writer visited the honey shows held in connection with the horticultural shows at Hucknall and Southwell and was struck with the variety of the honey shown. Evidently the NBKA has educated some of its members to a high standard and taught many of them how to produce good honey and also to place it on the exhibition table in a very attractive form. The entries were of such a high order at both shows, so that the work of the judge must have been a very difficult one indeed. A number of samples of fruit and hawthorn honey were staged at both shows, which are very dark in colour and of a very pungent flavour and tasted very strongly of hawthorn. We would give a hint here to intending exhibitors to show only the brightest and lightest coloured honey and really that obtained from the clover and stored in supers or sections. The nearer honey can be to a straw or pale amber colour, other things such as flavour, density, and aroma, being equal, the more chance you will have of winning a prize.

BBJ 13 Aug 1896
Leicestershire BBKA held its annual exhibition of honey in connection with the Leicester Horticultural Society, at the Abbey Park, Leicester, on August 4. The entries were the largest yet known, numbering upwards of eighty. The quality also was very fine, there being an evenness in the whole of the exhibits which made the competition very keen. A word of praise is certainly due to exhibitors in the "Honey Trophy Class" for the very tasteful way in which they were staged. Mr. Scattergood, jun., of Stapleford, Notts, officiated as judge.

BBJ 13 Aug 1896
The annual show of the Northants BBKA was held at Delapre Park, Northants, in connection with the Horticultural Exhibition, on August 3 and 4. The entries

numbered 192, nearly all being staged. Mr. FJ Cribb, of Retford, who officiated as judge, considered the exhibits made a very creditable display for a small association, both in quantity and quality.

BBJ 13 Aug 1896
Helmsley and District BKA (Yorks) The above recently-formed association held its first annual show of honey, etc, on August 6, in conjunction with the Ryedale Agricultural Society's Show in Duncombe Park, the beautiful seat of Earl Feversham. Mr. P. Scattergood. jun., of Stapleford, Notts, officiated as judge. In his remarks in the bee-tent he congratulated the Association on its first venture.

BBJ 20 Aug 1896
The Show at Chester. Twelve 1-lb Jars Extracted Honey — c, W. Herrod, Newark-on-Trent

NG 22 Aug 1896
NBKA Show. The general secretary of the NBKA sent us a reminder of the above the other day and also a schedule and entry form and we note from the same that this year the show is to be in connection with the Greasley, Selston and Eastwood Agricultural Societies' Show at Moorgreen on September 8th. There is a very interesting schedule and no less than eighteen classes for honey in various forms, honey vinegar, confectionery, bees, bees' wax, etc. The prizes offered are on a liberal scale.

BBJ 27 Aug 1896
Royal Lancashire Agricultural Show was held at Southport on July 31 and August 1 and 3, in fine weather. The honey department, though well filled in some classes, was scarcely up to the usual standard as far as number of entries. The quality of the exhibits, however, were excellent, some very fine specimens being staged. Mr. P. Scattergood, jun., Stapleford, Notts, staged a collection of articles of food, etc, in which bee-products figured as ingredients.

BBJ 27 Aug 1896
Bees Casting out Drones. Extracting from Brood Chambers.
Yesterday my bees were busy all day long throwing drones out of the hive, besides many like the enclosed, which I take it are drones ready to hatch out.
1. Is there anything unusual about this?
2. There are eight frames in the hive crowded with bees
3. Six are about a quarter full at top with honey, the seventh has a little brood, the rest honey; and the eighth no brood, but not quite filled with honey. Would you advise me to extract any honey? If not, will there be any necessity to feed next month?
W. Doleman, Nottingham, August 18.

Reply.

1. Throwing out drones and drone brood at this season is perfectly natural and need cause no alarm.
2. We strongly deprecate extracting from brood-combs in autumn and replacing natural food with syrup.
3. If the combs as described are left untouched, no feeding will be needed and this is by far the best course to follow, seeing that the trouble of extracting a few pounds of honey, only to replace it with syrup which requires labour on the part of the bees in evaporating moisture from the food and sealing it over, makes the whole operation a loss.

NG 29 Aug 1896
A number of questions have been asked as to the real difference in the colour of honey this year and the reason for it. It has been the writer's privilege during the last few weeks to visit shows in honey producing districts in Notts, Leics, Yorks, and Salop and there has been a marked difference in the honey staged. Some show tables filled with nothing but honey of a very dark colour, while others have had honey bright and light coloured with which to satisfy the most fastidious as regards taste. While we have a decided leaning to honey of a light colour, it must be distinctly understood that the flavour is a far more important point than colour. The reason for so much dark honey this year cannot be fully explained, excepting that the fruit and hawthorn bloom, especially the latter, produced a poor quantity of nectar this year and in those districts where the beekeeper has let the fruit and hawthorn honey be carried into the super and not extracted before the clover yield was being stored, the result has been a honey of a strong pungent flavour with poor density and a dark colour.

NG 5 Sep 1896
The Secretary of the British Dairy Farmers' Association has kindly sent a schedule of the forthcoming dairy show, to be held at the Royal Agricultural Hall, Islington, from the 20th to 23rd of October next. We would strongly advise our readers, especially those who belong to the NBKA to enter for one or more of the nine classes. We are inclined to think that the Notts. honey this year will take some beating. The members of NBKA have also an advantage owing to the association being associated with BBKA of getting half entry fee.

BBJ 10 Sep 1896
A correspondent sends us a cutting from The Field of August 29, referring to bees and certain lime flowers, which reads as under:
"I have read with interest of the bumblebees being found dead under lime trees in flower. We have at Rainworth a good many lime trees of various sizes; these generally have many flowers on them, and are frequented by bumble-bees, yet only under one lime are they found dead. On either side of the entrance gate are lime trees, which

BEEKEEPING IN VICTORIAN NOTTINGHAMSHIRE

form an arch covering the gate and carriage drive, and it is under one of these that every year great numbers of bumble-bees and a few hive-bees are seen lying dead; and, though these two trees are close together, I have never found one bee dead under the one on the left hand side of the gate, or under any of the other numerous limes dotted about the house and grounds." — J.Whitaker

"The note of Mr. F. Boyes is interesting, but he is undoubtedly wrong. These bees have not died under the ordinary lime tree, but they died in hundreds under the white lime trees (*Tilia petiolarts*) and as it was not at the end of summer, but just as the flower was in perfection. I think there can be no doubt that the honey poisons them. It seems to have no ill effect on any other insect. There have been, and still are, plenty of other flowers to support the bees, nor has there been any cold at night to injure them and, since the flowering of the limes is over, no more have died under them. Medway."

[We do not remember the question having been raised before, but there can be no doubt that the blossom of ordinary lime trees is in no way injurious to bees. Eds.]

Joseph Whitaker, JP, was born in 1850 at Ramsden Farm, Rainworth. He lived at Rainworth Lodge for mostof his life and died there in 1932. He was an acknowledged authority on birds and wrote several books on this subject. Amongst his traits was his habit of always walking on the road, never on the pavement.

The local secondary school is named after him.

Sheffield Daily Telegraph 11 Sep 1896
Derbyshire Agricultural Show. The Derbyshire BKA held their annual exhibition and the unfavourable season had an unfavourable effect on the entries. H Meakin of Nottingham was one of the winners.

NG 12 Sep 1896
The NBKA, with Mr A Warner acting as local secretary, also joined with the Agricultural and Horticultural Society after an interval of about five years and thus lent an additional attraction to the gathering at Moorgreen.

BBJ 17 Sep 1896
NBKA held their annual show at Moorgreen in conjunction with that of the Agricultural and Horticultural Society on the 8th inst. when over 900 lb of excellent honey was staged and proved a very attractive department of the show, the tent being crowded most of the day. Mr. TW Jones, of Etwall, Derby, was the judge, assisted by Mr. SW Marriott, of Nottingham, and made the following awards to NBKA members:
Best Hive—1st, GE. Puttergill, Beeston
Best Hive {made by Amateur)—1st, JT Faulconbridge, Bulwell; 2nd, JF Simpson, Underwood

Single 1-lb. Jar Extracted Honey (Open) - 3rd, T. Blake, Broughton; 4th, G. Marshall, Norwell

Honey Trophy—1st, J and W Herrod, Newark; 2nd, G. Marshall; 3rd, JWS Rawson, Selston; 4th, JT Faulconbridge

Twelve 1-lb. Jars Extracted Honey (Local) —1st, JWS Rawson; 2nd, Wm. Brooks, Eastwood; 3rd, GM Bolton; hc, J. Rawson, sen.

Twelve 1-lb. Jars Extracted Honey (Notts only).—1st, G. Marshall; 2nd, J and W Herrod; 3rd, P. Scattergood; 4th, H. Merryweather; 5th, GE Puttergill; hc, JF Simpson, CM. Lindley

Six 1-lb. Sections—1st, G Marshall; 2nd, J and W Herrod; 3rd, GE Puttergill; hc, JWS Rawson

Six 1-lb. Jars Granulated Honey—1st, JWS Rawson; 2nd, H. Wiggett; 3d, J and W Herrod

Shallow Frame of Comb Honey—1st, T. Marshall; 2nd, G Marshall; 3rd, J and W Herrod

Six 1-lb. Jars Extracted Honey—1st, GM. Bolton; 2nd, Geo. Hayes; 3rd, Wm. Poxon; 4th, J. McKinnon

Six 1-lb. Sections (Beginners only)— 1st, F. Wygett, Annesley; 2nd, G. Reeve, Moorgreen

Twelve 1-lb. Jars Extracted Honey (Blow's Jars)—1st, GE Puttergill

Twelve 1-lb. Sections (in Blow's Sections) — 1st, JWS Rawson

Honey Vinegar—1st, GE Puttergill; 2nd, P. Scattergood

Honey Cake—1st, P. Scattergood; 2nd, J and W Herrod

Observatory Hive—1st, H. Wigget; 2nd, J. Annable; 3rd, GE Puttergill; 4th, P. Scattergood

Wax — 1st, G. Marshall; 2nd, GE Puttergill; 3rd, J. Gray, Long Eaton

NG 19 Sep 1896

The principal event of the past week to beekeepers was the annual show of the NBKA which was held in connection with the Greasley, Eastwood, and Selston Combined Agricultural and Horticultural Societies' show at Moorgreen. The entries were numerous and varied and some very fine exhibits were staged. Mr. George Marshall of Norwell, is to be congratulated on his winning the silver medal, the bronze medal, and the bronze pendant, besides a large number of other classes. The honey trophy class attracted four exhibitors and the exhibit staged by J and W. Herrod, of Sutton-on-Trent, was generally admired and deservedly gained the first prize. Other rather novel features of the show were the classes for confectionery and vinegar, both made with honey. The recipe for making it was, too, attached to each exhibit.

Altogether this annual show is one of the best the Society has ever had and the new secretary, Mr. George Hayes and all his staff of willing workers are to be congratulated on their success. The staging and all the arrangements were all that could be desired. Mr TW Jones, of Etwall, officiated as judge and he was assisted by Mr. SW Marriott, of Nottingham. Several gentlemen asked for information as to the association and the terms of membership. All wishing for information on this subject should write to Mr. George Hayes, Mona-street. Beeston.

NG 26 Sep 1896

Honey Harvest in Notts. Now the season is over, it is well that we should review the same and compare notes. If any readers have anything good, bad, or indifferent to tell now is the time to tell it. I have seen reports from various parts of the county as to the quantity and quality of Notts. honey. A reader in Southwell says for comb honey it has not been a good year and the quantity has been small, sections only being moderately filled. Those who worked for run honey have had a fair harvest. The writer says his bees have averaged 40 lbs per hive.

Another correspondent says: At Newark we cannot boast of any large takes, but the honey is dark in colour. The takes in this district in frame hives average from 25 to 60 lbs per hive. A beekeeper at Bleasley has taken 340 lbs from five hives and his "Wells hive" has given 150 lbs. Early takes were very dark.

A writer in the north of the county says: "The honey crop here has been only moderate. Of course there have been a few good takes, but as a rule they have been moderate. The colour is not good, only fair and the density only fair."

In Arnold the honey has differed more than usual in colour and consistency. A nice quantity was gathered during the spring blooming period, but it was very dark and thick. Later on came the clover crop - much lighter but thinner.

In the Nottingham area the honey crop has been about on average and, as usual, some beekeepers have better results than others. I think that the honey is better than usual.

A large quantity of honey was staged at the annual show of the NBKA, at Moorgreen, on September 8th, was of very good quality and there was one thing in connection with the exhibits that was very noticeable. The Notts. beekeepers know how to stage their exhibits. Some exhibits which came from a distance contained particles of wax and pollen and in some cases dust. If our industry is to be successful all members of the "cult" must make the products of the hive, whether for sale or placed on the exhibition bench to look as nice as possible. Above all, they must be scrupulously clean and if this is so they would command a ready sale at good prices.

Minute Book – Quarterly Committee Meeting held at the Peoples' Hall, Heathcote-street, Nottingham on Saturday October 3rd, 1896. Present: Messrs Brooks, Gray, Herrod, Hallam, Faulconbridge, Forbes, McKinnon, Marriott, Pugh, Poxon, Puttergill,

Rawson, Scattergood, White, Warner, Wood and Hon. Secretary. Mr Scattergood being voted to the chair, the minutes of the previous meeting were read and confirmed.

A letter was received from his Lordship stating his inability to be present. A letter was read from Dr Sharp of Brant Broughton applying for expert and lectureship. After consideration it was resolved that the secretary informs Dr Sharp that his application could not be entertained.

A circular from BBKA about the County Honey Trophy for 1897 was read. It was resolved that the secretary answer be endorsed and that this association try and enter the competition and that the secretary ascertain how and when the honey for the Trophy be obtained and report to the next meeting.

The quarterly balance sheet was next read. Mr Herrod proposed and Mr Marriott seconded that the same be adopted.

Mr Pugh proposed and Mr Hayes seconded that a very hearty vote of thanks be given to Mr Warner for his help generally and for the staging in particular at the Moorgreen show.

The secretary asked whether gold pendants could be supplied in lieu of silver if the winner paid the difference in price. The committee agreed to this which was proposed by Mr McKinnon and seconded by Mr Wood.

NG 3 Oct 1896
JG. Long Eaton, writes: About six weeks ago the bees in my apiary commenced to rob a hive and I did not discover it for a few days owing to my apiary being a good distance away from home. Things got very bad and do what I could I could not stop it. The robbers completely emptied the hive and it was soon depopulated. I then closed the hive and let it stand for a fortnight in its original position empty. Last week I transferred another stock into it, removing the hive to the stand from which I transferred the bees and closed the entrance to half an inch. Strange to say the robbers commenced to rob the stock I had transferred, even though it was quite six yards away from its former position and in a very short time large numbers of bees were dead at the entrance.

I wish to ask first - What would you advise me to do? and second, Do you think the colour of the hive has anything to do with attracting the bees?

Evidently a bad case of robbery and I judge from your letter that you have found it a very difficult one to deal with, especially as the robbers have cleared one hive and have started of a second stock in the same hive but in another part of the apiary. As was stated a fortnight ago bees are by nature inveterate robbers and if not stopped the whole apiary may become demoralised. If what I advised on September 19th does not stop the trouble, then it may be advisable to close the entrances with a strip of perforated zinc and allow sufficient ventilation on the top of the frames and then remove the hive to a distant apiary for a week or two, if that is possible. If not then you may remove the hive and bees into a dark room (say a cellar) for a few days.

The strength of the individual bees will be conserved by the quiet and they will be better able to defend their hive afterwards. As there is no nectar in the fields for them to gather no loss is suffered by the confinement. The starting to rob another stock which was transferred to the same hive, although, as you say, the hive was removed six yards away and placed on the stand which the stock transferred to the hive originally occupied is certainly a strange proceeding and your question as to whether the colour of the hive has anything to do with attracting the bees, is a reasonable one under the circumstances. There is no doubt that bees do possess the sense of colour.

If your hive was blue probably this may have something to do with the robbers tackling it a second time, even though it was placed in a different part of the apiary. I would advise you to paint it white, and let me know if the robbers make a further attack.

CWH writes: I have been advised to get a flat top straw skep rather than a dome shaped top. Can you tell me where they may be purchased and what is the probable cost? I have tried in Nottingham and cannot obtain them. One dealer informed me they were not used now? Is this so?

Reply: Of course you do not intend to put any bees into a skep of any kind this season, because if they have not got the combs built and well stored before now it will be too late, but if you are looking ahead and wanting the information for next season I gladly help you. First then, the flat top skeps are more used now than ever where the old skep is still in use and you may obtain them for about 1s 6d or 2s each from most appliance dealers. If you cannot obtain them in Nottingham write to J. Howard, of Holme, Peterborough; C. Redshaw, Wigston; or W. P. Meadows, Syston; or try Abbott Bros., Southall (London). Be sure to say when ordering that you want a flat top skep. I bought several some time ago from a cottager in Cambridgeshire who makes them for the trade.

"North Notts." asks if there are two kinds of heather. He thinks he has seen two kinds and he is told that one produces honey and that the other does not. What is the difference, if any, and which is the honey producing kind?
Reply: Yes, there are two kinds of heather. The honey-producing kind 'Calluna vulgaris,' is usually known in England as "ling," and in Scotland as heather. The other

kind is known as bell heather. '*Erica cinera*,' is practically valueless to the beekeeper. Unfortunately in Notts, we have little or no heather excepting perhaps in the extreme north of the county.

NG 10 Oct 1896
I would like to draw attention to the fact that at a meeting of the RAS of England, at Manchester, in June next, it is arranged in connection with Bee and Honey Department of the Show to have a class for "County Honey Trophies," that is trophies from various counties, staged by the representatives of the Bee Keepers' Association, the same to contain approximately 300 lbs of honey in different forms. I am informed that NBKA have decided to enter the competition and to this end they are requesting members to keep in stock and lend or give the association a small sample, say six to twelve pounds of their best honey, of this or last years harvest. Of course, it is known that the early date of the show (June) precludes beekeepers from sending next season's honey, as the harvest in Notts will barely have commenced when the show is held. A special and urgent request is made to all members of the NBKA not to part with all their best honey. They will doubtless hear from the secretary shortly. We should all like our county not only to be represented, but to be successful. The suggested prizes are substantial and are worth competing for.

BBJ 22 Oct 1896
Wintering of Bees. Attic vs. Cellar.
 In the autumn of 1887 I had between fifty and sixty hives of bees to pack away for the winter months and amongst them a lot of badly-made hives, which, in spite of all my efforts, would get damp inside. At the time I was living in a veritable paradise for bees i.e., Nottinghamshire and close by the Dukeries.
I had read many varied experiences of wintering bees, as described by correspondents in the BBJ, and it was from a Scotch bee-keeper's experience that I had read which induced me to try a rather unusual experiment. I possessed a very large and roomy attic with a concrete floor and into this attic in the middle of November I transported twelve hives of bees. I boarded the window up and otherwise made the place as "dark as a bag," as the saying goes. I had also three large dry cellars. The cellar steps entered into No. 1. cellar; from No. 1 you entered No. 2 by a doorway and from No. 2 to No. 3 the same. Into No. 3 cellar I also transferred twelve colonies. I watched my experiments for the first week or two with considerable trepidation but all went on smoothly and I began to shake hands and congratulate myself on having at last found an ideal method of wintering bees.

But alas! I had a rude awakening. There had been a fall of snow during the earlier part of January and about the middle of the month there came a glorious hot day which dispelled the snow like magic. A few days afterwards I thought I would go and see how my attic friends were getting on. My first footstep into the darkened attic told me, "like a flash of lightning," the result. I was walking on a thick carpet of dead bees. I let daylight into that attic in double quick time, only to find that the whole of my twelve colonies carpeted the room. The glaring hot sun had raised the temperature of the attic and also of the bees, which had left their hives to perish on the floor.

Although my attic speculation had turned out a miserable failure, I am glad to record that my cellar experiment was a complete success. I don't think I lost a hundred bees in the cellar and the majority of these perished at different times, whilst I was examining with a lighted candle.

Unfortunately, I didn't weigh my stocks at the time of taking them into the cellar, nor when I brought them out at spring and so cannot give exact information as to amount of stores consumed; but I know it was very little. As near as I could estimate the consumption was not more than 6 lb per hive. In conclusion, I may state the winter packing of my attic friends varied considerably from that adopted with my cellar customers and although I blamed old "Sol" at the time as the cause of my attic disaster, later experiments have taught me that to the difference of packing my non-success was mainly due. John Goodall, Doshill Lodge, near Tamworth

NG 31 Oct 1896
As I stated a fortnight ago a contingent of Notts bee men visited the Dairy Show at Islington, last week and among them the writer and certainly this year's was the best exhibition of honey, etc. that has ever been held in connection with the British Dairy Farmers' Show. The arrangements both for staging and affect left nothing to be desired. There were 158 entries for honey, etc, from 25 English counties, and 11 from Scotland. Our own county of Notts, had 4 entries and 1 prize.

BBJ 5 Nov 1896
The last BBKA Conversazione of the present year was held at four o'clock on the 22nd ult. at Jermyn-street, St. James'. The spacious board-room of the RSPCA was unusually crowded by an audience consisting largely of bee-keepers who had been brought to London by the attractions of the Dairy Show at the Agricultural Hall, whence they were invited to attend this meeting. Mr. Henry Jonas, Hon. Treasurer BBKA was voted to the Chair and the Council of the Association was represented by (amongst others) AG Pugh (Notts). Among the general visitors will be found hon. officials and prominent members of the following County Association (amongst others) Notts, including G. Hayes and P. Scattergood.

NG 5 Dec 1896
JMK Epperstone, writes, "Winter passages, or bee ways to enable our little friends free access from one frame to another without having to either go round by the cold hive floor or remain on the centre combs and die of hunger, with, perhaps, plenty of

stores in the outer frames. I most certainly endorse your advice that a good large piece of candy placed on the top of the frames is the best and most advantageous winter passage.

I give you herewith my method (for what it is worth) of placing candy on the hive. I make a shallow box, 12 inches by 6 inches and $2\frac{1}{2}$ inches in depth, all outside measurements. The top is after the same principles as Howards rapid syrup feeders, ie. a sliding glass and wood cover over. For the bottom I take ordinary builders laths used for plastering and cut them in lengths same as the box. The lath I divide into three in the width and nail the strips on the bottom of the box, leaving a space between strips of about three-eighths of an inch; then I place the box in the top of the frames so that the strips will be at right angles to the tops of the frames in the hive.

The quilts must be removed before placing the candy box on the hive, so that as many frames as possible will be covered by the bottom of the candy box. The candy may of course, be placed in before or after, just as is most convenient, then all is covered over snug and warm. The advantages of this box for candy feeding will be obvious. There will be a perfect bee passage underneath the feeder, then with the aid of the glass cover, the least peep, without any loss of heat, will satisfy the beekeeper as to how the supply of candy is holding It out and no bees will be lost in the manipulation, there will be no sticky mess quilts and no inconvenience caused by bees building brace combs above. "

The candy feeder described by our correspondent is certainly very ingenious and I have no doubt answers well. The only candy box or feeder I use, however, is a wood section holder with the glass at the top and into these I run my candy and the bees soon clear it out and thus make passage ways for themselves. I have had a peep at three of my hives to-day (November 27th) and find that during the last month they have each consumed nearly $4\frac{1}{2}$ lb of candy and I shall early next week place another cake on the hives to renew those nearly consumed. I must thank my correspondent for his letter and also for the description of his ingenious candy box and hope more of our readers will let us hear from them on any matter of interest in the apiary.

Minute Book – Quarterly Committee Meeting held at the Peoples' Hall on Saturday, January 2nd 1897 with Mr Scattergood in the chair. Present: Messrs Annable, Gray, Forbes, Faulconbridge, Herrod, McKinnon, Pugh, Poxon, Rawson, Turner, White and hon. Secretary.

The minutes of the previous meeting were read and confirmed.

The secretary then read a letter from his Lordship stating his inability to be present.

BEEKEEPING IN VICTORIAN NOTTINGHAMSHIRE

The quarterly balance sheet was read and it was proposed by Mr Pugh, seconded by Mr Herrod that, in future, all prize money for honey and appearing in our schedule be paid by NBKA. Mr Herrod proposed and Mr Gray seconded that the balance sheet with the alteration as pointed out be adopted and printed with a notice convening the annual meeting and sent to each member.

It was proposed that the annual meeting be held in the Peoples' Hall at 3pm on Saturday February 20th, that a tea be provided and arranged by the secretary on the same lines as the previous year at 1/- each and that Mr Meadows be asked to give a lecture or lead the converszione, we paying his railway fare.

It was arranged that the following should be given for the prize drawing:
1 hive   1 smoker   250 labels   1 vol Records   1 Guide book and 100 labels

County Honey Trophy. This subject was discussed and ordered to be put on the agenda for annual meeting.

BBJ 14 Jan 1897
My Christmas Holiday. Receiving a pressing invitation from a brother and sisters, living in the county of Notts, to spend a few days with them at Christmas, I prevailed upon "the good wife" to join me in accepting the invite, and on December 24 we started for the famous district known as "The Dukeries." My brother, well knowing my love for "the bees" took care that I should not fail to have an opportunity for interchanging ideas with a few bee-keepers located in the district and of holding one of those friendly chats about our hobby which bee-men never fail to enjoy. It was rather late on Christmas Eve when we reached Newark and the evening was employed in other than bee-talk.
The morning of Christmas day was beautifully fine, the sun shining bright and warm as in spring. We were early among the bees and I had a good inspection of one apiary where the bees were flying quite freely and strong on the wing. I availed myself of an offer to look into the hives and was very pleased to be of some service in consequence; for although the bees were in perfect health some stocks were found dangerously short of stores. One queen showed her prolificness and good breeding powers by having already nearly filled a whole comb with brood and eggs.

To see the bees of a good queen like this wanting the needful stimulus of a "full cupboard" was more than I could stand and the owner of the bees, not being very well versed in making candy of the right sort for winter feeding, I volunteered to make some on the spot. Unfortunately there was no cream of tartar in the house, but, not to be put off in so critical a case of impending starvation, I started accompanied by our friend's son as a guide, for Newark, where, with the help of a friendly chemist, we got our cream of tartar and returned satisfied.

It was not long before we got the sugar and "stewpan" in operation, with myself as chief cook and you would have been amused to see yours truly - with shirt-sleeves rolled up - at work over a big fire, making - not the pudding - but a Christmas feed for the bees! I explained the "why" of each part of the process of making soft candy and our friend, while looking on, said, his candy turned out " stringy "enough, but, as I explained to him, it was on the "cooling off" properly that so much of success depended and this "cooling off" meant constant stirring - with the pan immersed in cold water - until the mass got stiffish and turned to something like the consistency and colour of honey that was nearly granulated white.

When the candy had cooled sufficiently and was of the proper consistency, while still warm, I showed in practice how food given that way stirred into activity a starving stock on a fine day in December and made the bees safe for many weeks to come. Having had ocular demonstration of the way it worked, our friend thanked me for the lesson. And later in the day I was introduced to another beekeeper, then a third proving how readily "birds of a feather flock together."

Bees and bee-talk so entirely occupied us that we had, very considerately, assigned to us a separate room all to ourselves and there with experiences - jovial and otherwise - and all round hearty enjoyment, we sat till the " Sma' hours ayont the twal," as our Scotch friends say. The bee flora of this part of Notts. consists mainly of clover in the meadows and willows (*Salix viminalis* and *S. triandra*). The pollen laden catkins of these two species of osier willows have a very stimulating effect on bees in the early spring. I know this from personal knowledge of my own district in Hunts.

Altogether, I was very pleased with the surroundings of the whole district comprised in the famed Dukeries from the bee-keeper's point of view and I hope to make another journey to the same neighbourhood in the coming summer.

After a pleasant time in the interim, I travelled on Monday the 28th to a place near Mansfield, whereto I had sent a stock of bees in the spring. They had gathered enough stores to keep themselves, but no surplus. I therefore concluded that the district was not a good one for bees. This was proved later by our driving over to visit a bee-keeper who has sixteen stocks at a small railway station nearby and where I was told the surplus harvested in '96 did not average 20 lb per hive. Friend Pugh, of Beeston, will understand this, as he has one stock "boarded out" in this same apiary. On Thursday, the 31st, we drove to Pleasley and saw there an apiary of twelve stocks. Unfortunately, however, there was no one at home, so no information was obtainable either about bees or the district as a honey-producer.

I was quite delighted with the beautiful scenery about here. Friday - New Year's Day - saw us on the road to visit Newstead Abbey and during the journey we came across some heather, which we don't get in Hunts. On Saturday we started for home after a most enjoyable tour. The final conclusion I arrived at was:

Well, our county may be flat and wanting in the beauty of landscape we had left behind, but, with all its shortcomings, give me old North Hunts for bee-keeping. Wishing all bee-keepers a prosperous honey season in 1897.

Richard Brown, Flora Apiary, Somersham, Hunts.

Apart from its condescending nature this story loses veracity when we are led to believe that a chemist in Newark was open on Christmas Day!

Minute Book – Annual General Meeting held in the Peoples' Hall, Heathcote-street, Nottingham on Saturday, February 20th, 1897. Owing to the unavoidable absence of the president, Viscount St. Vincent, the Rev HL Williams was voted to the chair. The secretary read the minutes of the last annual meeting and these were confirmed.

The annual balance sheet was next read and after some remarks and explanation of the same it was ordered that the same be adopted and printed with the annual report. Proposed by Mr Scattergood and seconded by Mr Pugh.

The annual report was then read and again some discussion on the same it was moved by Mr Turner and seconded by Mr Warner that the same be adopted and printed.

Mr Hayes moved and Mr Wright seconded "that a hearty vote of thanks be accorded his Lordship for his services as president during the past year and also for the substantial help he has given to this association and that he be asked to remain our president for the ensuing year."

Mr Scattergood then moved and Mr Herrod seconded that as Mr Pugh had resigned the office of treasurer, that both offices of treasurer and secretary be merged into one. Mr Scattergood moved and Mr T Marshall seconded that Mr Hayes be re-elected secretary and treasurer for the ensuing year. Mr Rawson moved and Mr Pugh seconded that Mr Scattergood be thanked for his services as auditor and that he be re-elected for the present year. It was proposed that the committee be elected '*en bloc*' with the exception of Mr Merryweather of Southwell who was unable to get to the meetings and that Mr Pugh be elected to fill this place. It was next proposed that Messrs Pugh and Hayes be representatives to the BBKA.

It had been proposed in committee by Mr Raven and seconded by Mr Rawson that Mr Pugh be made a life member in consideration of his services as secretary and this was brought to this meeting to decide. Mr Scattergood opposed this and after some discussion Mr Pugh asked that the question be allowed to drop.

Mr Gray moved that the wording of Rule 4 be altered, but as there was no seconder the proposal fell.

An adjournment was then made for tea to which about 60 members and guests sat down and at about 6pm the meeting was resumed.

Mr AG Pugh having been voted to the chair, proceeded to award the medals and certificates won at the Annual County Show - silver and bronze medals of the BBKA and Notts. silver pendant to Mr G Marshall of Norwell, Notts. silver pendants Mr W Herrod and Mr T Marshall, Sutton-on-Trent and certificates to Mr J Sopp, Mr J Rawson, Mr GE Puttergill and Mr P Scattergood.

The special class for a 1 lb bottle of honey was then judged. Mr A Warner (Moorgreen) gained first and Mr Scattergood, jun., (Stapleford) second. The prizes were kindly given by Mr Meadows of Syston and the honey (24 bottles) was given to the Children's Hospital, Nottingham. Mr AE Newton of Lowdham offered a special prize of 10/- for the best jar of fruit preserved in honey to be brought to the next annual meeting.

Mr Meadows of Syston then spoke on the evolution of appliances, pointing out their best uses and his remarks were altogether very interesting and instructive.

Mr Scattergood then moved and Mr McKinnon seconded that this county do enter the County Honey Trophy competition at Manchester in June 1897. Mr Scattergood moved and Mr Herrod seconded that the details to be left to the committee.

Mr Scattergood proposed and Mr Riley seconded that the secretary write to the Notts. Agricultural Society to receive a deputation to consider the advisability of joining them with the annual honey shows.

It was next arranged that a class for dark honey should be included in the annual show and that the secretary should affix to the schedule coloured papers shewing what was meant by light and dark honey. Mr Herrod proposed and Mr Clark seconded that we have a class for a 1 lb section on the same lines as the 1 lb bottle, open to all comers. It was resolved that the winner of the prize for amateur hive be debarred for one season.

The meeting was brought to a close by the annual prize drawing which resulted as follows:

1st	WBC hive	given by the President	won by Mr McKinnon
2nd	WBC hive	given by the President	won by Mr Jones,
3rd	WBC hive	given by the President	won by Mr Scrimshaw
4th	Swarm catcher	given by Mr Meadows	won by Mr Knowles

BEEKEEPING IN VICTORIAN NOTTINGHAMSHIRE

5th	Rapid feeder	won by Mr Glew
6th	Brigham smoker	won by Mr Raven
7th	250 honey labels	won by Mr Hayes
8th	Vol 'Records'	won by Mr Frampton
9th	Guide book	won by Mr Ball
10	100 honey labels	won by Mr Clark
11th	New floor feeder	won by Mr Measures

NG 21 Feb 1897

The annual meeting of the members of this Association was held on February 20th, 1897, at the People's Hall, Heathcote-street, Nottingham, the Rev. HL Williams (Bleasby) in the chair. Amongst those present were: Messrs. Geo. Hayes (secretary), P. Scattergood (auditor), AG Pugh, RW Turner, J. Herrod, W. Herrod, T. Marshall, G. Marshall, J. Wright, G. Wood, A. Warner, J. Rawson, W. Poxon, J. Gray, WP Meadows (Syston), RG Glew, M. Lindley, J. Annable, and others. The Secretary read a letter from Viscount St. Vincent (President) expressing his regret at being unable to attend and preside upon this occasion.

In moving the adoption of the balance sheet, which showed a small deficiency, Mr. Scattergood said he had gone carefully through the accounts and could bear testimony to the accuracy with which they had been kept. The Secretary announced that Viscount St. Vincent had sent a cheque for £6, with a request that the balance, after payment of his subscription of one guinea, should go towards making up the deficiency on the balance sheet. Mr. Pugh, in seconding the motion, said it behoved them to see if there were not more Lord St. Vincent's in the county. It was announced that the Chairman, the Rev. HL Williams, had doubled his yearly subscription.

The balance-sheet was unanimously adopted.

The Secretary then read the annual report, which dealt with the bee-season of 1896 and the work of the past year generally. Among other items it was stated that there was a small increase of membership compared with that of '95. Also that the Notts County Council had increased their grant for technical instruction in bee-keeping from £20 in '95 to £30 last year. It was also hoped that a further increase would be got for 1897. The report was adopted unanimously and a vote of thanks to Viscount St. Vincent for his services to the Association and a request that his lordship be asked to again accept the position of president, was carried *nem. con.*

Mr. George Hayes was elected treasurer and secretary to the Association and Mr. P. Scattergood re-appointed auditor. Messrs. Hayes and Pugh were chosen as representatives on the BBKA.

After which several matters of interest were discussed, including a proposed entry for the County Honey Trophy Competition at the Royal Show at Manchester.

The meeting concluded with a vote of thanks to the chairman for presiding.

At the close of the meeting tea was partaken of by the members and a distribution of medals and certificates followed. The meeting terminated with the usual prize drawing.

BBJ 25 Mar 1897
The first quarterly BBKA Conversazione for the year 1897 was held at six o'clock on March 12, in the Board-room of the RSPCA Jermyn-street, St. James'. Mr. Till occupied the chair and there was a good attendance. Among those present was P. Scattergood. The Chairman expressed regret that Mr. P. Scattergood had been obliged to leave to keep an appointment, as the members of the BBKA took special interest in the work taken up by the gentlemen named in his own district and which commended itself to him as being very useful.

Minute Book – Quarterly Committee Meeting of NBKA held in the Peoples' Hall, Heathcote-street, Nottingham on Saturday April 3rd, 1897. Present: Mr P Scattergood in the chair, Messrs Annable, Faulconbridge, Gray, Herrod, Hallam, Marriott, Pugh, Puttergill, Rawson, Turner and hon. Secretary.

The minutes of the last meeting were read and confirmed.

The secretary then read a letter from the hospital thanking them for the honey, a circular from the BBKA about Foul Brood Bill and a circular on foul brood statistics, a letter from the Notts. Agricultural Society.

Messrs Marriott, Turner, Scattergood and Hayes were, at this point, deputed to meet the NAS committee, the meeting being adjourned for half an hour for this purpose. When they returned they reported that the NAS offered £10, we find the tent and staging and they have the entrance fee. After considerable discussion, it was proposed by Mr Scattergood and seconded by Mr McKinnon that if the NAS will transpose their offer to £15 and they to find tent and staging and have the entrance fees and gate - we will accept and find a judge, lecturer and bee tent.

The following gentlemen were asked to form a committee for special business in relation to the Honey Trophy competition, Messrs Hayes, Herrod, Pugh and Scattergood.
As the day was far spent it was decided to again adjourn the meeting to a date to be decided upon by the secretary.

BEEKEEPING IN VICTORIAN NOTTINGHAMSHIRE

BBJ 8 Apr 1897
WANTED, copy of report with names and addresses of Members of the Lincolnshire, Yorkshire, and other Bee Associations. Walton & Co., Muskham, Newark.

Minute Book – Special Committee Meeting held at the Peoples' Hall, Heathcote-street, Nottingham on Saturday, May 1st, 1897. Present: Mr P Scattergood in the chair, Messrs Annable, Brooks, Faulconbridge, Forbes, Marriott, Pugh, Rawson, Turner, White, Newton and hon. Secretary.

The minutes of the last meeting having been read and confirmed, it was proposed by Mr Brooks and seconded by Mr Annable that we join Hucknall with a show this year on the terms of previous years. A letter was read from the secretary of the Horticultural Society at Lowdham stating that they did not want us to join them this year. Mr Scattergood proposed and Mr Faulconbridge seconded that we join Moorgreen this year as in former years.

A letter was read from the Norwell Society stating that they objected to NBKA paying prize money. Mr Marriott proposed and Mr Scattergood seconded that under these conditions we do not join them.

A letter was read from the Southwell Society asking for a slight alteration in the schedule which was agreed to and it was then decided we should join them on the usual terms and that the secretary be asked to write and try to get the last class made into one for those not having previously taken a prize.

A letter was then read from the secretary of the NAS stating they accepted our proposals.

The schedule was carefully gone through and after sundry minor revisions the following was agreed to:
 "give the Silver Medal to light honey class
 give the Bronze Medal to the dark class
 to add a class for competition in bee driving and to give a 10/-
 prize each day to the winner
 that Mr Scattergood be assistant to the judge each day and for
 these services be paid 7/6d
 that Mr Turner be judge at Moorgreen
 that Mr Pugh be judge at Southwell."

BBJ 3 Jun 1897
June 9 and 10 at Nottingham. In connection with the Notts Agricultural Society, at Colwick Park. Shows in connection with the NBKA will be held as under:

Hucknall Torkard, July 20. Entries close July 16;
Southwell, July 22. Entries close July 5;
and Moorgreen, September 7. Entries close August 27.

BBJ 10 Jun 1897
Bottles (Cardiff). Feeding Bottles.
There is no need for trouble as to these. If you cannot procure the one known as the "Nottingham," buy an ordinary regulating feeder from some of our advertisers, any of which will answer the purpose.

BBJ 24 Jun 1897
The annual show of NBKA was held in connection with the show of the County Agricultural Society in Colwick Park, Nottingham, on June 9 and 10. The exhibits were of good quality and the staging and arrangements in the show-tent reflected great credit on the energetic secretary of NBKA, Mr. Geo. Hayes and his staff of willing workers.

Notts bee-keepers are very enthusiastic showmen and work with commendable energy to make their exhibitions successful. This year the County Agricultural Society, as representing the larger industry, has generously helped the prize fund and, by so doing, secured what has long been hoped for by the Beekeepers' Association, viz. the holding of the annual shows of both societies at the same time and place.

Lectures were given in the bee-tent with suitable demonstrations and manipulations during each day by Mr. WB. Webster, of Binfield, Berks, who also officiated as judge, and made the following awards:
Amateur-made Hive—1st, J. McKinnon, Gedling; 2nd, Geo. Hayes, Beeston
Single 1-lb. Jar of Extracted Honey (open) —2nd, J & W Herrod, Sutton-on-Trent; 3rd, AG Pugh, Beeston
Single 1-lb Section (open) — 2nd, J & W Herrod
Honey Trophy—1st, J & W. Herrod; 2nd, JW Rawson, Selston
Twelve 1-lb Jars Extracted Honey (light)—1st, P. Scattergood, jun., Stapleford; 2nd, H. Merryweather, Southwell; 3rd, J & W. Herrod
Twelve 1-lb Jars Extracted Honey (dark) — 1st, W. Hallam, Orston; 2nd, J & W. Herrod; 3rd, JW Rawson
Six 1-lb Sections — Equal 1st, J & W Herrod and G. Marshall, Norwell
Six 1-lb Jars Granulated Honey—1st, J & W Herrod; 2nd, AG Pugh; 3rd, H. Merryweather
Shallow Frame of Comb Honey—1st, P Scattergood, jun; 2nd, J & W Herrod
Extracted Honey (Novices only)—1st, W Hallam, Orston
Honey Vinegar—1st, P. Scattergood, jun. 2nd, J & W Herrod
Honey Cake—1st, J. Wilson, N. Clifton; 2nd, P. Scattergood, jun.

BEEKEEPING IN VICTORIAN NOTTINGHAMSHIRE

Bees in Observatory Hive — 1st, JW, Rawson; 2nd, G. Hayes
Beeswax—1st, J. Wilson; 2nd, W. Hallam, 3rd, P. Scattergood, jun.

BBJ 1 Jul 1897
The Fifty-eighth Annual Meeting of the Premier Agricultural Society of England took place at Trafford Park, Manchester, in the Diamond Jubilee week, amidst splendid weather and the most favourable surroundings. The main attraction of the Show was, of course, the Special County Trophy class, in which nine exhibits were staged. In this class the prizes were exceptionally liberal; more valuable, we believe, than at any previous bee exhibition held in this country. Amongst the counties represented was Nottingham.

Coming to the first prize Trophy, i.e., Notts, it formed a very pretty display, arranged in the form of a square pyramid, so placed on the table -space that the corners of the pyramid came in centre as it faced the onlooker and left a clear space at the corners of the table, on which were arranged four small groups or pyramids of sections and jars, four tiers high, each surmounted by a vase of flowers. The main or centre portion was six tiers high, each tier being supported by stages of beveled plate glass, while the apex of the whole was surmounted by a tasteful arrangement of flags and flowers, intermingled with an abundance of maidenhair ferns. The foundation of the stand was draped with lace and flags. The contents of the whole stand was as follows:
Sixty-four 1-lb sections, two shallow frames of comb-honey, about 200 lb of extracted honey in various-sized screw-cap jars, about 14 lb wax (in cakes and large moulds), ½ gallon mead and $\frac{1}{4}$ gallon vinegar. The total weight of honey was about 298 lb.

Special County Honey Trophy Competition.
Class 375. Best and most attractive display of Comb and Extracted Honey, and such

Honey products as Wax, Mead, and Vinegar, arranged in Trophy form on a space not exceeding 4 ft. 6 in. square, by 5 ft. in height. The gross weight of the Honey (which may be in any form and of any year) must approximate 300 lb (12 Entries).
First prize (£15 and silver medal), NBKA
Class 385. Bees-wax—Not under 3 lb (17 entries) — c, J. Wilton, Newark.
Class 387. Honey Vinegar (6 entries) — 3rd, P. Scattergood, Stapleford, Notts.

This silver medal now forms the President's badge of office worn by successive holders for the three years of their appointment.

In the photograph of the winning display, George Hayes, secretary of NBKA, is on the left wearing his characteristic bowler hat and William Herrod is on the left with his equally distinctive mustache.

Minute Book – Quarterly Committee Meeting held in the Peoples' Hall, Heathcote-street, Nottingham at 3pm on Saturday, July 3rd, 1897. Present: Messrs. Brooks, Forbes, Faulconbridge, Gray, Hallam, Merryweather, Pugh, Puttergill, Rawson, Scattergood, Turner and hon. Secretary.

Mr Pugh having been voted to the chair, the minutes of the previous meeting were read and confirmed. The quarterly balance sheet was read and discussed. Mr Scattergood proposed and Mr Merryweather seconded that the same be received and adopted.

Correspondence was then read which had reference to the minutes and also an apology from his Lordship for non-attendance with congratulations on our success in winning first prize in the County Trophy Competition at Manchester Royal Show.

Mr Scattergood next moved that the secretary be asked to write to the Notts. Agricultural Society, tendering our thanks for the grant and staging at Colwick and make overtures for a show with them next year.

It was next ordered that the secretary should be empowered to spend a reasonable sum in advertising in the Grocers' Journal or other such paper as he may think fit and also if possible appoint local agents for the sale of Notts. honey, preference being given to Messrs J Barber and Sons.

Some conversation here ensued as to the bee-driving competition at the annual show and as the first days driving was not altogether satisfactory, Mr Pugh (who was one of the competitors) did not wish for any prize but as Mr Herrod (another competitor) suffered the loss of a queen, it was decided to award him half the prize (5/-).
After congratulating each other and thanking the sub-committee of the Trophy, the meeting terminated.

BEEKEEPING IN VICTORIAN NOTTINGHAMSHIRE

BBJ 8 Jul 1897

A Protest about the County Trophy Class.

I feel, after the stand I myself, along with others interested in the County Trophy class, took on the show-ground in protesting against the decision of the judges, that some further explanation is due to both parties interested (I mean the judges and the exhibitors). The Notts. exhibit was found to be nearly 20%, under the requirements of the schedule and the Yorks exhibit was 20%, over the prescribed limit—a full 40% of difference between the two.

I give due credit to the Yorks. staging committee that they honestly marked down their quantity; but when we consider those in charge of the Notts. trophy certifying that their stand contained a larger quantity of honey than it actually did, they must either have been lax in their counting or they certified to a deliberate lie, to screen themselves, with an insufficient quantity.

Wm. Woodley, Beedon, Newbury, June 26

BBJ 15 Jul 1897

Judging at the Royal Show.

As a member of the Council of the BBKA, upon which body lies full responsibility with regard to the appointment of judges and others who hold office at shows held by the Association, permit me to say that never before in my experience have I seen what I regard as so complete a misappropriation of awards as was seen in the Trophy Class at the 'Royal' last month.

I am also well acquainted with three of the four judges, but, along with practically all the other exhibitors, including the representative of the Notts Association, was never more surprised in my life than when the awards were made known, so difficult was it to believe that any one of the judges could be guilty of what was generally regarded as a great error of judgment.

The question of the Notts. and Yorks. trophies not being in accordance with the regulations contained in schedule is a matter which rests entirely with the officials of the Show, whose duty it is to see that every exhibit is in conformity with the schedule, and does not affect the judges at all.

Henry W. Brice, Vale Park, Upper Norwood, July 9.

BBJ 15 Jul 1897

I write from Notts. and along with Mr. Hooker (p. 262), regret exceedingly the remarks in your editorial and review of the trophy class in the issue of BBJ of July 1 and shall I say that, in common with a number of leading Notts. beemen, I think they are a "wee bit" one-sided. Their tone shows a little disappointment on your part as one of the judge and, to say the least of it, the article is none too friendly to our Notts. trophy. On this point many of us think we have just cause for complaint, especially in view of the fact stated by Mr. Hooker, that "honey alone was not to decide the merits of the trophy." I could say much more on this point, but refrain, as in the main I think it is a matter for those who made the awards and not for me, to defend their action.

But, when we come to the "exhibition of temper" (for I cannot call it anything else) shown by Mr. W. Woodley, not only in the BBJ, but also on the show ground when the awards were made known, it is a different matter and I must say I was much surprised and pained and, but for the credit of our Notts. trophy, and our Association and the integrity of the Committee, I would have passed over such an outburst of egotism and bad temper with the contempt it deserves.

Sirs, we in "Notts." are not guilty of deliberate lying and the Committee who had charge of the trophy, of whom I am one (and without egotism, please), are all men of honour and integrity quite as much so as Mr. Woodley; while our Secretary is not, as a rule, lax in anything. I do not intend to satisfy Mr. Woodley at present as to the actual weight of the trophy. I will say this, however, that he is a long way out in his percentage of what he pleases to call " the requirements of the schedule"; but I suppose this is the point at issue and one reads 'approximate' to mean one thing and Mr. W. another.

I am afraid that Mr. W. is a bad loser and a disappointed man; while his exhibit - disqualified several years ago - which appears to stick, has no bearing whatever on the case in point. A trophy exhibit and a section exhibit are under quite different conditions and his remarks, re perversion of judgment, are uncalled for and ungentlemanly and I must say I gave Mr. W. credit for having more manliness and common sense. His sneer, too, about a "handful of maidenhair fern" being accepted before his "cream of cream", "quality of honey" and the "grandest display," etc., is in keeping with the outburst of egotism and temper before named and are hardly worth consideration.

Mr. W. and those who were dissatisfied had their remedy. They might, if they had so wished, lodged a protest and done the matter in a straightforward, honourable and manly form; but, instead of this, he, Mr. W., rushes into print and calls on judges and committee to disqualify the Notts. and Yorks. exhibits and, of course, place his (Berks) exhibit first. He well knows this is too late now, but if it could be done, it might satisfy a spoilt child and that child Mr. Wm. Woodley.

P. Scattergood, Junr, Stapleford, Notts, July 10.

Lincolnshire Echo 15 Jul 1897
Lincolnshire Agricultural Show. Granulated Honey – 3rd J. Cribb, Retford.

BBJ 17 Jun 1897
Trade Catalogues Received.
EC Walton & Co., Muskham Works, Newark.—60 pages. Besides a very full list of bee goods to suit all classes of bee-keepers, Messrs. Walton & Co. devote some

16 pages to poultry houses and the various appliances connected therewith. It may interest poultry breeders to know that Mr. T. Gascoigne, after twenty-five years as poultryman with various gentlemen—who during that time took nearly all before them in the show pen with the various breeds they kept—has now connected himself with the firm of Messrs. Walton & Co. Poultry-keepers living in the neighbourhood who may be requiring advice, will therefore do well to make Mr. Gascoigne's acquaintance. We are informed that he intends taking up judging appointments.

BBJ 17 Jun 1897
To Be Sold, Handsome Oak Observatory hive. Cost £7. Will take £2. Address. Miss Cane, Aveston Rectory, Newark.

BBJ 1 Jul 1897
Joiners Wanted, Bee Appliances. Constant work to good men winter and summer. Walton, Muskham, Newark.

BBJ 15 Jul 1897
The monthly meeting of the BBKA Council was held at 105, Jermyn-street, the 9th inst., Mr. Till occupying the chair. Two new members were elected: Mr. Geo. Hayes, Mona-street, Beeston; Mr. Wm. Herrod, Sutton-on-Trent.

BBJ 29 Jul 1897
The NBKA show at Hucknall on the 20th inst. in connection with the Horticultural Society, was a very fair one, some splendid honey in sections and bottles being staged. The first prize extracted honey was of excellent quality. Mr. P. Scattergood (first-class expert) officiated as judge and also gave some well attended and interesting demonstrative lectures in the bee tent. The following is the list of awards:
Twelve 1-lb. Sections — 1st, J & W Herrod, Sutton-on-Trent
Twelve 1-lb. Jars Extracted Honey—1st, H. Wiggett, Hucknall; 2nd, Mrs. Hind, Papplewick Grange; 3rd, JT Faulconbridge, Bulwell Wood; 4th, H. Cartledge, Hucknall
Bees in Observatory Hive—1st, G. Hayes, Beeston; 2nd, H. Wiggett; 3rd, H. Merryweather, Southwell

BBJ 29 Jul 1897
The Southwell Show was held in connection with the Horticultural Society, at Southwell on the 22nd inst. The entries in all six classes were numerous, and the general quality of the honey staged excellent, especially the sections. Mr. P. Scattergood (in the absence of Mr. Pugh) officiated as judge, and his awards were as follows:
Shallow Frame of Comb — 1st, W. Lee, Southwell; 2nd, J & W Herrod, Sutton-on-Trent; 3rd, T. Marshall, Sutton-on-Trent
Six 1-lb Jars Granulated Honey—1st, W. Lee; 2nd, H. Merryweather, Southwell; 3rd, G. Wood, Apperstone

Six 1-lb Sections—1st, J & W Herrod; 2nd, T. Marshall; 3rd, H. Merryweather

Six 1-lb Jars Extracted Honey—1st, JT Faulconbridge, Bulwell; 2nd, H. Merryweather; 3rd, G. Marshall, Norwell; 4th, W. Lee

Bees in Observatory Hive—1st, G. Hayes, Beeston; 2nd, G. Marshall; 3rd, W. Lee; 4th, J & W Herrod

Single 1-lb Jar of Extracted Honey {local} —1st, H. Merryweather; 2nd, W. Lee; 3rd, J. Holmes, Southwell

BBJ 19 Aug 1897
An Unrehearsed Incident.
At the invitation of his Grace the Duke of Portland, Messrs. P. Scattergood, jun., and AG Pugh visited the show of the Welbeck Tenants' Agricultural Society, annually held in the grounds of Welbeck Abbey, to lecture on bee-keeping and to give demonstrations with the bees. The Society was established eight years ago for the tenants on the various estates belonging to the Duke of Portland and the annual show has become a very popular one, being looked forward to by many besides tenants with great pleasure. The meeting under notice took place on the 3rd inst. and was no exception to the rule, as seen in the fact that over ten thousand persons visited the show ground during the day.

When the bee-tent had been set up a rather novel incident happened, as follows:
The bees intended for use in driving and manipulating had been conveyed from Nottingham in a skep and the weather being very hot at the time, they were liberated as quickly as convenient after arrival on the ground; instead, however, of settling down quietly the bees swarmed. This put the two lecturers in a bit of a fix, seeing that the Duke and Duchess and the house party from the Abbey had intimated their desire to hear the lectures and witness the manipulations with the bees.

By the time the distinguished visitors had arrived at the tent the swarm had nicely settled on the branch of a lime tree about thirty yards away and so, after explaining the position of affairs, it was suggested that the party should proceed to the spot and watch the hiving of a swarm. This was agreed to and skep in hand Mr. Scattergood mounted a box and had soon successfully hived the swarm, amidst the applause of those present, numbering several hundreds. The bees were then thrown down in front of the frame-hive and soon ran in, thus affording a practical object-lesson in housing a swarm in a frame-hive, all of which details were explained to the company.

This is the first time that representatives of the NBKA have been invited to take a part in the proceedings at this show and judging from the large numbers gathered round the bee-tent at the three subsequent lectures much good and many new members must accrue—at least, we hope so.

BEEKEEPING IN VICTORIAN NOTTINGHAMSHIRE

BBJ 26 Aug 1897
The Shropshire BKA's annual show and honey fair was held, as usual, in connection with the Shropshire Horticultural Society's Fete in the famous 'Quarry' at Shrewsbury on August 18 and 19 and was in every respect a conspicuous success. Amongst the judges was Mr. P. Scattergood, jun., Stapleford, Notts. who judged the extracted honey.

BBJ 26 Aug 1897
Staffordshire BKA show - Amongst the prizewinners were J and W Herrod – 3rd, 12 1lb sections.

Leicester Chronicle 18 Sep 1897
Loughborough Agricultural Show. Short lectures on the management of bees were given by Mr P Scattergood.

BBJ 23 Sep 1897
Leicestershire BKA held its third annual exhibition of honey, etc, in connection with the Loughborough Agricultural Society, on September 15, in the beautiful grounds of Southfield Park, kindly lent by WB Paget, Esq. The modern methods of manipulating bees were shown by Mr. P. Scattergood, Stapleford, Notts. who also acted as judge.

BBJ 30 Sep 1897
Colonist (Nottingham). Packing Hives for Transit Abroad.
Our correspondent will find full details on the subject of his inquiry in the June issue of our monthly, the Record, which may be had from this office for $2^1/_2$d in stamps. We have no knowledge of bee-keeping in South America beyond the fact of knowing that large imports of honey and wax are received here from that quarter of the world.

Minute Book – Quarterly Committee Meeting held in the Peoples' Hall, Heathcote-street. Nottingham on October 9th, 1897. Present: Messrs. Annable, Brooks, Forbes, Faulconbridge, Gray, Hallam, Newton, Marriott, Mackender, Pugh, Puttergill, Rawson, Scattergood, Turner, Warner, White and hon. Secretary. Mr Mackender having been voted to the chair, the minutes of the previous meeting were read. Mr Pugh proposed and Mr Scattergood seconded that the same be received and adopted. The secretary then read correspondence relating to aforesaid minutes and it was ordered that the secretary should try still further and if possible lead to the appointment of an agent for Notts. honey bearing our labels.

The quarterly balance sheet was next read and discussed, when Mr Scattergood proposed and Mr Pugh seconded that the same be received as read.

The secretary then read a list of awards for the 1897 shows, all of which had been duly paid.

The Trophy account was next gone into and it was agreed to give Mr Herrod a gold medal for his assistance at Manchester.

It was then stated that we should most likely require more labels printing next season and it was considered advisable to try and get a smaller one if possible. The secretary is to prepare samples for the annual meeting.

BBJ 28 Oct 1897
BBKA Conversazione. At the commencement of the proceedings, the board-room of the RSPCA was crowded by an audience which included bee-keepers from nearly all parts of the kingdom, among those present were Geo. Hayes, AG Pugh and P. Scattergood, jun.

BBJ 28 Oct 1897
The British Dairy Farmers' Association held their twenty-second annual exhibition of stock, produce, and implements in the Royal Agricultural Hall, Islington, the show opening on Tuesday, October 19th and closing on Friday, 22nd.
Twelve 1-lb Jars Granulated Honey (13 entries) — vhc. and reserve No., H. Merryweather, Southwell; CW Lee, Southwell.

Minute Book – Quarterly Committee Meeting held in the Peoples' Hall, Heathcote-street, Nottingham on Saturday January 8th, 1898. Present: Messrs. Brooks, Faulconbridge, Forbes, Marriott, McKinnon, Poxon, Pugh, Puttergill, Turner, Scattergood, Rawson, Warner, Gray, Herrod and hon. Secretary. In the absence of Lord St. Vincent, Mr Marriott proposed and Mr McKinnon seconded that Mr Pugh be asked to take the chair. Mr Pugh accepted and the minutes of the previous meeting were read and dealt with serratim when Mr Scattergood proposed and Mr Marriott seconded that the same be confirmed by the signature of the chairman.

The quarterly account was next gone into after which Mr Gray proposed and Mr Turner seconded that the same be received and adopted.

The secretary next read the annual balance sheet and after discussion thereon, Mr Scattergood, hon. Auditor, stated that he had carefully examined this account with all vouchers accounted for and as he believed it was a faithful statement proposed that the said be adopted and printed with the circular convening the annual meeting and sent to each member. Mr Herrod seconded this and it was carried unanimously.

The arrangements for the next annual meeting were next gone into. Mr McKinnon moved and Mr Scattergood seconded that the annual meeting be held on Saturday, February 26th at 3pm prompt and also that the Mayor and Sheriff (Frederick F Gregory) be asked to support his Lordship in the chair. Mr Marriott proposed and Mr McKinnon seconded that a meat tea be provided at 4.30pm and that the secretary be asked to provide this on similar lines as on previous occasions. This he consented to do.

The secretary then stated that, with the assistance of a few friends, he had been able to arrange for the following competitions at the annual meeting:

		1st	2nd	3rd
Class 1	Best single jar or bottle of fruit preserved in honey	7/d	5/-	4 cakes candy
Class 2	Best sample of honey wine or mead to approx 1 pint	7/d	5/-	4 cakes candy
Class 3	Best cake made with honey in lieu of sugar, made by exhibitor or his wife	7/d	5/-	4 cakes candy
Class 4	Best single bottle 1897 granulated honey Honey for Children's Hospital	2/d	1/-	
Class 5	Best sample of honey vinegar to approx 1 pint	Certificate		

Mr Turner proposed and Mr Marriott seconded that Mr Meadows be asked to come and judge each of the above classes. This was carried and Mr Pugh kindly undertook to pay his expenses if he charged them.

The question of the annual prize drawing was next considered and it was arranged to obtain the following if his Lordship did not send anything.
1 wax extractor   250 honey labels   1 volume 'Records'   Guide book   100 labels

Mr Hayes promised a solar wax extractor, a super of shallow frames fitted with 'Weed' foundation. Mr Puttergill offered a rapid feeder and Mr McKinnon a candy feeder.

There is some confusion here. Does the Minute book indicate both dignitaries were present? Frederick W Gregory was Sheriff of Nottingham in 1898/1899. The Mayor of Nottingham from 1896 – 1899 was Edward Henry Fraser.

BBJ 13 Jan 1898
Homes of the Honeybee. If there is anything in the principle of heredity it will explain why Mr. P. Scattergood, Jun. whose apiary is seen, together with himself and wife, in the picture below - is to-day one of our most ardent bee-keepers. Fifty or sixty years ago his grandfather kept bees and was considered very successful with them, while some twenty-five or thirty years ago an uncle owned a large number of hives in North Notts, having at one time had an apiary of 240 colonies in frame-hives and skeps.

Mr. Scattergood is, we believe, the oldest member of the Notts BKA, which he joined at its formation in April, 1884 and the only one remaining of those who became members at the first meeting, held on April 30 of that year. His bee-keeping dates a little earlier than this; and, having caught the bee-fever, he began making his own hives, engaging in 'driving' expeditions and, as he says, made the error, so common to beginners, of having more stocks than he could manage. In this way and through various causes - coupled with the severe winter of '87 - he lost all his twelve hives, gained some useful experience and, nothing daunted, began again. This time he went more steadily to work and his present apiary is the outcome.

Writing us as to his present view, he says, "I have tried various kinds and makes of hives during the last thirteen years, buying, making and selling again many styles and types of hives, gaining experience as I went on; but I like the "WBC" hive best of all and gradually all my present hives will give place to this kind." The hives seen are by various makers, while several - including the "Wells" - are home made.
"As will be seen, the apiary is situated on a hill-side near my house and is arranged in terraces, but the picture, which could only be got from a bed-room window, hardly gives a correct idea of the contour of the ground; the top of the hill being about seventy feet higher than the roadway."

We learn, also, that the garden is fully planted with choice fruit-trees; while among many beautiful flowers grown in profusion in various beds, the rose evidently finds a large place in the garden and in the heart of its owner, upwards of 400 trees of various varieties finding a place on the "hill-side." Nearly all these fruit and rose trees have, we believe, been raised, budded, or grafted by our friend himself, who never tires of the enjoyment afforded by garden and bees.

Mr. S. further says, "The district is only a moderate one for bee-forage, but honey of very good quality is sometimes obtained." He also adds, "I sell all my own honey and at times a good deal for friends. All is sold locally and retail without difficulty, but I never sell any but the best quality and, being made-up attractively, customers come year after year."

BEEKEEPING IN VICTORIAN NOTTINGHAMSHIRE

Our friend is in the prime of manhood, is forty-two years of age and has a good wife and a happy home. Full of buoyant energy, he is a busy man and holds several responsible appointments, among others being those of Clerk to the School Board of his native town, Secretary and Manager to the Waterworks Company, an Overseer of the Poor and a prominent member of the Parish Council. He is also a local preacher and Sunday school teacher. Though having no family of his own, he is a dear lover of the young and a general favourite with the young people. Full of schemes to help his fellow-men, he is always willing as a beekeeper to impart instruction and to give help to all who need it. A diligent student of natural history, he possesses a library on bees and bee-keeping and kindred subjects of which any man might be proud.

In conclusion, it must be added that Mr. Scattergood is a familiar figure at the Quarterly Conversaziones of the BBKA in London; nor does he grudge a long journey to town in the "Dairy Show" week. Holding the 1st class certificate of the Association, he has done good service in lecturing on bee-culture in connection with Technical Education and frequently takes charge of the bee-tent at Shows, where his ready speech and hearty manner are much appreciated by listeners who are onlookers. We hope his activity among the bees may long continue.

NG 15 Jan 1898
The County Beekeepers' Association.
Through the kindness and courtesy of the secretary of the above association I was favoured last week with a perusal of the yearly balance-sheet of the association. The wonder to me is how so much work is covered and prizes given at so little cost of working expenses. The association commenced the year with a balance owing to treasurer of over £7, but by rigid economy, coupled with the success of the county in winning the first prize in the "Honey Trophy" competition at the Royal, the association has a balance of £1 16s to its credit to start the new year with. The amount given in prizes during the year at the various shows has been £24 5s 6d and the total income from all sources has been £75 19s 1d. The subscriptions from members has been £29 10s. What to me is the only discouraging feature in the balance-sheet is that the association is so meagrely supported by those who might help them. I find that out of a total membership of 147, there are only nine persons who subscribe over 10s each. The secretary informs me that that they are a diminishing quantity. This ought not to be. Surely in our own county there are men of wealth and position, who could easily spare, say, a guinea per annum, to support an industry like beekeeping, and to help the association to assist more than it can possibly now do those whose slender incomes do not allow them largely to support a Beekeepers' Association.

NG 10 Feb 1898
The general secretary has sent me a circular of the annual tea and meeting of NBKA which is to be held in the Peoples' Hall, Heathcote-street, Nottingham on Saturday,

February 26th. The president, Viscount St Vincent, is announced to preside and in the evening a converszione is to be held, when a number of interesting matters are to be introduced for discussion. In looking through the balance-sheet and cash account I find that while the association has lost some of its supporters, who gave liberally to its funds, the members have more than paid their way, mainly owing the their success in winning the first prize with the trophy at Manchester and also the liberal donations of the president and others. I hope to meet a number of my correspondents there, besides many leading men of the craft.

Minute Book – Annual General Meeting held in the Peoples Hall, Heathcote-street, Nottingham on Saturday, February 26th, 1898.

The secretary at the outset read a telegram from Lord St. Vincent, stating his inability to be present, whereupon Mr Scattergood was voted to the chair.

The minutes of the previous meeting having been read and confirmed – the secretary next presented the annual balance sheet and after consideration Mr Pugh proposed and Mr Scattergood seconded that the same be adopted and printed with the annual report.

The secretary then read the annual report after which Mr Pugh proposed and Mr McKinnon seconded that the same be adopted and printed.

The secretary next moved that a hearty vote of thanks be accorded Lord St. Vincent for his presidency during the past year and that he be asked to accept the same for the coming year. Carried with acclamation.
Mr Turner proposed and Mr T Marshall seconded that Mr Hayes be re-appointed treasurer and secretary for the coming year. Mr Gary proposed and Mr Pugh seconded that Mr Scattergood be re-elected hon. Auditor for the year. Mr Rawson proposed and Mr Lindley seconded that Messrs Hayes and Pugh be re-elected representatives to the BBKA meetings. The committee were re-elected 'en bloc' with the exception of Messrs Annable and White resigned. The following gentlemen were also elected to the committee – Messrs Wilson, Harrison and the Rev CWH Griffith.

The secretary read a letter received from Mr GH Young of Nottingham stating that his bees had been upset and some thrown down a well and asked if this association could help him in bringing the offenders to justice. Mr Pugh proposed that bills be printed offering the sum of one pound reward to those who should give evidence to the conviction of the offenders.

Proposed by Mr Ellis and seconded by Mr Gray that the members of the sub-committee for technical instruction be elected annually and that their names be printed in the annual report.

An adjournment was made for tea to which 70 members and friends sat down. At 6.15pm the meeting was resumed.

Mr Scattergood who still occupied the chair first proceeded to award the prizes won in the competitions arranged for this meeting which were as follows:

Class 1 For best single jar or bottle of fruit preserved in honey
  1. Mr Smith, Bradmore
  2. Mr Scattergood, Stapleford
  3. Mr Hayes, Beeston

Class 2 For best sample of honey wine or mead to approx. 1 pint
  1. Mr Raven, Nottingham
  2. Mr Raven, Nottingham
  3. Mr Maskery, Kirkby

Class 3 For best cake made with honey in lieu of sugar, made by exhibitor or his wife
  1. Mr Scattergood
  2. Mr Warner, Moorgreen
  3. Mr Poxon, Nottingham

Class 4 For the best single bottle 1897 granulated honey.
  1. Mr Bolton, Eastwood
      Honey for Children's Hospital
  2. Mr Scattergood

Class 5 For best sample of honey vinegar to approx 1 pint
  Mr Scattergood

Objects of interest. New inventions, etc. were next shewn and discussed, these included Simmins Conqueror hive, Sladens slit sections and foundation fixer, no-beeway sections and dividers, metal and cleated, etc.

Mr Herrod of Sutton-on-Trent next shewed some slides of Mr Howards foundation factory and apiary and of the Manchester Trophies, etc.

The question as to the disposal of the Trophy Medal caused an animated discussion and it was proposed by Mr Pugh and seconded by Mr Gray that a case or frame be provided to take both card and medal and that the same be kept in the custody of the secretary and exhibited at shows and other such places to the honour and advancement of the association.

It was next ordered that as soon as the present stock of labels is exhausted that the secretary procure another supply of a same pattern, but of a smaller size so that the rate of the new label be 1/- per 100, 4/6d for 500, and 8/6d for 1000, postage free.

A very successful meeting was brought to a close with the annual prize drawing which resulted as follows:

1st	Wells Hive	given by President	won by Mr Puttergill, Beeston
2nd	Hive	given by President	won by Mr Rawson, Selston
3rd	Solar wax extractor and feeder	given by Mr Hayes	won by Mr Forbes, Radcliffe
4th	Super of shallow frames	given by Mr Hayes	won by Mr Lindley, Beauvale
5th	Rapid feeder	given by Mr Puttergill	won by Mr Wadsworth, Newark
6th	250 honey labels		won by Mr Henshaw, Lambley
7th	Candy feeder	given by Mr McKinnon	won by Mr Marriott, Nottingham
8th	1 volume 'Records'		won by Mr Scott, Nottingham
9th	Uncapping knife and brush	given by Mr Raven	won by Mr Fox, Kirkby
10th	Guide book		won by Mr Henton, Norwell
11th	Wax smelter	given by Mr Raven	won by Mr Harrison, Nottingham
12th	100 labels		won by Mr Mills, Newark

Rev Clement William Haslewood Griffith was rector of All Saint's church, Winthorpe 1895-1918

BBJ 3 Mar 1898
The annual meeting of the members of NBKA was held on February 26, 1898, at the People's Hall, Heathcote-street, Nottingham. In the absence of the president (Viscount St. Vincent), who telegraphed from London expressing his regret at being unable to attend, Mr. P. Scattergood, jun., was voted to the chair.

The balance-sheet having been presented, the Chairman, in moving its adoption, said the accounts reflected great credit upon their hon. secretary, who was the right man in the right place. They had more than paid their way during the year. They began the year with an adverse balance due to the treasurer and they finished with a balance on the right side.

The motion having been agreed to, the Hon. Secretary (Mr. George Hayes) read his report for the year, in which, after reviewing the past honey season in Notts, reference was made to the success of the association in winning the premier prize in the County Honey Trophy Competition at the 'Royal' Show in June last. This had given an impetus to the sale of Notts. honey, a large quantity of which had been sold for members. Among other satisfactory items in the report, it stated that the Notts. County Council had renewed their grant for education in bee-keeping and lectures had been arranged for at various centres in the county, and it was hoped the Council would see their way clear to increase this grant, for in no other department were they getting so much work done for so little money. Thanks were due to all who so materially assisted during the year.

After some remarks by the chairman, the report was, on the motion of Mr. Pugh, seconded by Sergeant Mackinnon, adopted.

The following gentlemen were appointed to serve on the executive committee for the year 1898: Messrs. W. Brooks, JT Faulconbridge, C. Forbes, SW Marriott, W. Poxon, AG Pugh, GE Puttergill, RJ Turner, Wilson, TN Harrison, J. Hardy and the Rev CWH Griffith.

A vote of thanks was accorded to Viscount St. Vincent for his services as president and it was resolved to ask his Lordship to continue in office for another year. Other officers were re-appointed.

After an adjournment for tea, the meeting resolved itself into a conversazione. Prizes were awarded for specimens of fruit preserved in honey, honey-wine, honey-cake, honey vinegar and pure honey. The successful competitors included Messrs. J. Smith, P. Scattergood, G. Hayes, T. Maskery, and GN Bolton, Mrs. Scattergood, Mrs. Warren, and Mrs. Raven.

NG 5 Mar 1898
I accepted the invitation to the annual meeting of NBKA at the People's Hall in this city on Saturday, and was soon made to feel at home receiving a hearty beekeepers' welcome from the genial and hard-working secretary, Mr. Geo. Hayes and also from other leading members of the association. Unfortunately, the noble president at the last moment was prevented from attending the meeting and this I could tell was a great disappointment to those present. The association is fortunate in having for a president a practical beekeeper like Lord St Vincent. His cheery presence to say nothing of his financial assistance, must be a tower of strength to it. The association has had a successful year and go out with a small balance in hand but what is needed is more assistance from those who can afford to give liberally and if this were so more work could be done and better results obtained.

The secretary's report grasped the situation and placed the year's work before the members in a clear manner. A matter of some interest was mentioned. It appears that a beekeeper residing in the city has been the victim of a cruel outrage. A number of his hives have been thrown over and ruined, while another has been thrown down a well. My only wish is that the offenders may be brought to justice. The association has done the correct thing in supporting its members by offering a reward for evidence that will lead to a conviction.

NG 5 Mar 1898
The annual general meeting in connection with the NBKA held on Saturday afternoon at the People's Hall, Heathcote-street, Nottingham. In the absence of the president (Lord St. Vincent) the chair was occupied by Mr. P. Scattergood, jun. There was a

numerous attendance of members.

The annual report presented by the secretary stated: The last season had not been for beekeepers that which was expected until just before the honey flow. The early spring was all that could be desired: stocks strengthened up rapidly up and by the end of May were ready for good work, but the flow of nectar in many districts was somewhat restricted owing to the cold nights which were experienced when the clover was in bloom. Some districts had certainly been more favoured than others but, from the information to hand the honey harvest in the county, as a whole, had only been about a second-rate one.

At the last annual meeting we decided to enter for the county honey trophy competition at the Royal Show, held at Manchester. It was gratifying to record that the association succeeded in carrying off the premier prize, viz. £15 and the silver medal, much to the chagrin of some of the other competitors. This ought to and, indeed, had given an impetus to the sale of Nottinghamshire honey - a benefit that would accrue to each member, directly or indirectly. It was a Jubilee honour, gained on Jubilee day.

The secretary was glad to state that last season he had been able to dispose of about 900 lb of honey and a quantity of wax for the members and this, he thought they would agree, was another step in the right direction. He should be glad, whenever the opportunity offered, to put those who had a surplus of honey for disposal in communication with buyers, if they would only make use of the association's free exchange mart.

The hon. auditor of the association, Mr P Scattergood, jun., had gained a 'First Class' certificate as expert. Thus for the first time in its history the association had the honour of possessing a first class expert.

With regard to the membership of the association, the present strength was 161, an increase of 13 as compared with the previous year. The association had undoubtedly taken a wise step in joining their annual show with that of the Notts. Agricultural Society, not from a monetary view, because the NAS did not offer them an adequate sum, but from the fact that they were joining hand in hand with agriculture, a branch of which beekeeping was avowed to be. And it was for this reason they were prepared to sink a little, hoping that the council of the NAS would increase their grant. The honey staged at last year's show was a decided success. Mr Scattergood obtained the BBKA silver medal for light-coloured honey and Mr Hallam of Orston, the bronze medal for dark.

The statement of accounts showed that receipts during the year had amounted to £75 9s 4d and the expenditure to £73 13s 4d leaving a balance in hand of £1 16s. The treasurer explained that the subscriptions had totalled up to £29 10s, as compared with £30 3s 6d, in the preceding year, whilst the donations had amounted to £28 12s as against £29 3s 2d in 1896. On the receipts side was an item of £7 17s 10d which represented the balance from the trophy fund. This amount had materially assisted the finances of the association. With regard to the expenditure, the cost of printing had been reduced from £11 10s in 1896 to £6 16s 6d last year. The reports were adopted.

Lord St. Vincent was requested to again accept the presidency; Mr. G. Hayes was re-elected hon. Secretary and treasurer; Mr. P. Scattergood, jun. was re-appointed auditor. Messrs. Pugh and Hayes were re-elected delegates to the BBKA.

The committee was appointed:
The Rev. CWH Griffiths, Messrs. W. Brooks, JT Faulconbridge, C. Forbes, SW Marriott, J. McKinnon. W. Poxon, AG Pugh. GE Puttergill, RJ Turner, P. Scattergood, W. Wilson, T. Harrison and J. Hardy.

Other business of a routine character having been transacted, the meeting terminated and the members partook of tea. Subsequently a conversazione was held, when the following subjects were discussed: The disposal of the trophy medal, a fixed price for honey, a new honey label and the disposal of honey. Various competitions took place, a portion of the honey exhibited being afterwards sent to the Children's Hospital. The usual prize drawing was also held.

BBJ 10 Mar 1898
The annual meeting of the Leicestershire BKA was held on March 5th, at the Victoria Coffee House, Leicester, under the presidency of the Mayor (Ald. Wakerley). There was a good attendance of members, among those present being Mr. and Mrs. Pugh, and Mr. Geo. Hayes, Sec, NBKA.

Minute Book – Committee Meeting held in the Peoples' Hall, Heathcote-street, Nottingham on Saturday, April 6th, 1898 at 3pm. Present: Rev. Griffith, Messrs Brooks, Faulconbridge, Forbes, Ellis, Hallam, Hardy, Harrison, Gray, Puttergill, Pugh, Rawson, Scattergood, Marriott, Wilson, Warner and hon. Secretary.

Mr Pugh having been voted to the chair, the minutes of the previous meeting were read. Mr Scattergood proposed and Mr Hardy seconded that the same be received and adopted. The quarterly account was next gone through by the secretary in detail, Mr Scattergood proposed and Mr Ellis seconded that the same be received.

The appointment of the sub-committee for technical instruction by the aid of the Notts. County Council grant was the next business and Mr Marriott proposed and Mr

Gray seconded that the number of members should be increased from six to eight. Mr Harrison proposed and Mr Ellis seconded that the existing committee be elected 'en bloc'. Mr Scattergood proposed and Mr Pugh seconded that Mr Ellis be one of the additional members. Mr Marriott proposed and Mr Warner seconded that Mr Turner be the other additional member.

Mr Marriott proposed and Mr Scattergood seconded that we accept the invitation to join the show at Moorgreen. Mr Harrison proposed and Mr Hardy seconded that we accept the invitation to join the show at Southwell and that the secretary endeavour to get the last class made into one for amateurs.

The schedule for the annual show at Colwick was next gone through in detail and the secretary was instructed to make the necessary alterations therein and print and send out to members as early as possible.

The secretary was instructed to try and arrange a show at Sutton-in-Ashfield.

BBJ 21 April 1898
Honey and Increase in One Season.
I have two strong stocks of bees in skeps, but wish to have a try working with frame hives. Now, supposing that each skep throws off both a swarm and a cast (which I hope they will, because of my desire to increase as much as possible), I ask:
1. Can I work the swarms and casts (two of each) into four strong stocks before the winter, supposing it is a good season, and I feed them well?
2. If the above works out as desired, will the parent stocks be likely to fill any supers?
3. I intend to drive the bees out of the skeps after the honey-flow is over and unite the two lots in a frame hive and feed up to make a fifth stock, unless you tell me a better method?
4. I suppose it would not do to put supers on before the honey flow, as this would tend to prevent swarming?
As a beginner with bees, I should be grateful for your opinion on the above.
North Notts., Retford, April 8.
Reply.
1. If you are fortunate enough to secure two early swarms, get the young queens safely mated and if the season turns out a good one there need be no difficulty in working them into strong stocks as desired; but don't overlook the "ifs" in making your calculations.
2. No, it is unreasonable, as a rule, to expect surplus from a skep that has thrown off two swarms.
3. The wisest course would be to use the bees and young queens from the parent skeps for strengthening the top swarms in the first established frame-hives, as these latter will be all the better for having young queens at their head for next

year's work after removing the old ones.
4. No; if you expect too much - in the way of both honey and increase - in one season, disappointment will probably follow.

BBJ 28 Apr 1898
We wish to exhibit (not particularly to compete for prizes) bee-appliances at the Royal Show. But on reference to the schedule, we notice that Class 341 reads the same as last and many previous years, viz. "To consist of the following articles," and lower down follows the words, "no articles must be added to the collection." We would be glad if you would give us some explanation of this clause, because last year and in many previous years we have noticed that none of the exhibitors have adhered to this rule and yet have not been disqualified. If the above rule is to be ignored (as it was last year) and exhibitors allowed to bring an ironmonger's shop, a tinware shop, a printer's shop, etc,, we should be glad to know, so that we might take staging for a few cottages, greenhouses, portable buildings, etc, in addition to the specified bee appliances (?) mentioned in rules.
EC Walton & Co., Muskham, Newark, April 23
[The only "explanation" we can offer is to remind our correspondents that the words quoted must be read along with the context. The full clause reads thus: *"No articles must be added to the collection, nor any portion of the exhibit removed during the show."* (The italics are ours.) We might also add - by way of further explanation - that in the paragraph specifying the appliances which must be included in the collection there follows the words: "And other distinct articles at the discretion of the exhibitor." It goes without saying, however, that in a class for "Collection of Hives and Appliances," only such things as are used in bee-keeping are admissible. Eds.]
BBJ 28 Apr 1898
Two Strong Stocks of Bees, in frame hives, with accessories. Hamel, Devonshire Promenade, Lenton, Nottingham.

Sigismund Hamel was born in Hamburg in 1824. In the 1881 Census he was living in the Nottingham Castle area.

NG 7 May 1898
I have received a copy of the schedule of prizes offered at the annual show of the NBKA which I am glad to find is again being held in connection with the Notts. Agricultural Society's' show in Colwick on June 1st and 2nd, as well as schedules for shows later on. There are prizes offered for appliances, honey in various forms, honey products such as vinegar, honey cakes, beeswax, etc. and also for bees in observatory hives. The schedule also states that there will be an examination for third-class certificates and that the bee tent will also be in attendance at the show. Entries close May 25th. The show is early but is to be hoped that a good number of members and others will make entries and thus make the show a success.

BBJ 26 May 1898
Bee Shows to come. June 1 and 2 at Colwick Park, Nottingham. NBKA in conjunction with the Notts. Agricultural Society. Liberal prizes and medals for bees, honey, and appliances. Schedules from George Hayes, Hon. Sec. Notts. BKA, Mona-street, Beeston.

NG 4 Jun 1898
The annual exhibition of the NBKA was held in a special tent on the ground of the County Agricultural Show at Colwick Park, the stewards being Messrs. Marriott (Sneinton), Puttergill (Beeston), and McKinnon (Carlton) and the judges, Messrs. CN White (St. Neots) and R. Turner (Radcliffe). For the condition of the season the exhibition was a very successful one, samples being sent from all parts of the county. Owing to the time of the year none of this season's honey could be shown. The show tent was rendered very attractive by reason of the decorations and there were staged two trophies each containing 80 lbs of honey. During the day Mr. P. Scattergood and Mr. CN White, together with Mr. McKinnon, of Carlton and Mr. Lee, of Southwell, gave exhibitions explaining the methods of driving and taking honey and contrasted the old system of skep hives with that of the bar frame. Mr. Scattergood, the well-known expert, lectured at brief intervals and demonstrated the various ways of manipulating bees. One of the most important points of his lectures was the emphasis he laid on the advantages of bee-keeping to agriculturists. By way of illustration he quoted an experiment of Darwin, who protected 20 heads of clover to prevent contact by bees. Those heads failed to produce a single seed, whereas 20 unprotected heads produced over 2,900 seeds, thus showing that the fertilising powers of the bee were of advantage to the farmer.

NG 11 Jun 1898
Many of our readers like myself visited the Nottingham Agricultural Show at Colwick last week. Although the climatic conditions were enough to damp the ardour of any but enthusiasts, there was a large number of the "cult" present on both days. The exhibits, considering that they had to be held over from last year, were fairly numerous and of excellent quality. The granulated honey class was well filled and some excellent honey was staged. Only three bottles of the current year's honey were staged and it was evidently gathered from the fruit and hawthorn blossoms. In the class for run or extracted honey some very good samples were staged. The competition in this class was very close. The four samples of wax were of good quality, especially that which gained the first prize. The samples of honey vinegar were an interesting class, showing how honey may be used in producing as excellent article for use at the table and for the preparation of salads. Many people are able to take honey vinegar who cannot take the ordinary kind and find no ill effects whatever. Another advantage is that it does not stain either table linen or silver.

BEEKEEPING IN VICTORIAN NOTTINGHAMSHIRE

The appliance exhibit as staged by Mr. GH Varty of Etwall was up to the mark and contained many articles of great utility for the beekeeper. Although only one exhibit was staged it gained the first prize. In Class 2, for the best and most complete hive for general use unpainted and made by the exhibitor or his *bona fide* workmen Mr. McKinnon, a member of the NBKA deservedly won premier honours and all who saw and examined the careful and accurate workmanship were pleased with the award. The observatory hives were a continued source of interest to visitors and the secretary and experts and others interested in the show and in charge of the tent, were at all times during the two days willing to explain the mysteries of the hives to visitors, to show the different kinds of bees, viz. queen, drones and workers and to give explanations to all who wished of the natural history, as well as the metamorphosis or changes through which the bees pass from egg to imago. Many of these quiet talks, were greatly appreciated and caused additional interest in the lectures and demonstrations which were given in the bee tent by the lecturers during the two days. Mr. Hayes, the general secretary of the association, worked very hard indeed to ensure the success of the show and he was ably seconded by his staff of willing assistants. From our own observations we are quite sure the assooiation is doing good work and is deserving of more financial support from those who can afford to help. If this were forthcoming I am assured that "bee talks" on village greens and in beekeepers' gardens could be arranged. Thus the cottager and labourers might be helped in a more practical way than is now possible owing to the lack of funds.

BBJ 16 Jun 1898
Wanted, A Young Man who understands bees and appliances to help at shows and make himself generally useful. EC Walton, Muskham, Newark.

BBJ 23 Jun 1898
Royal Agricultural Society of England. Birmingham Meeting, 1898. The fifty-ninth annual meeting of the Royal Agricultural Society opened at Birmingham on the 20th inst., under the most favourable weather conditions it is possible to imagine.
Classs 341. Collection of Hives and Appliances—3rd, EC Walton & Co., Muskham, Newark
Class 359. Honey Vinegar—1st and 2nd, Peter Scattergood, Jun., Stapleford, Notts.

BBJ 30 Jun 1898
Good Healthy Swarms, 2s 6d per lb. Herrod, Trentside Apiary, Sutton-on-Trent, Newark.

Minute Book – Quarterly Committee Meeting held in the Peoples' Hall, Heathcote-street, Nottingham on Saturday, July 2nd, 1898, at 3pm. Present: Messrs Brooks, Forbes, Ellis, Faulconbridge, Hardy, Jones, Marriott, Puttergill, Scattergood, Turner,

Warner and hon. secretary.

Mr Scattergood having been voted to the chair, the minutes of the previous meeting were read and on the motion of Mr Brooks, seconded by Mr Forbes, the same were confirmed and adopted.

The correspondence was next read including apologies for absence from several members of the committee.

The quarterly account was next gone through by the secretary when Mr Marriott moved and Mr Warner seconded that the same be passed as satisfactory.

Mr Puttergill proposed and Mr Brooks seconded that Mr Scattergood be the judge at Moorgreen show. An amendment by Mr Rawson seconded by Mr Forbes that Mr Turner should officiate. The proposition was carried. Mr Marriott proposed and Mr Faulconbridge seconded that Mr Pugh officiate as judge at Southwell show.

The application from the Swinderby Show was next considered and it was proposed by Mr Ellis and seconded by Mr Warner that under our present circumstances it was advisable not to make a grant and especially as it was not in the county of Notts.

A letter from his Lordship was discussed and it was considered advisable to await his Lordships' further expression on the matter.

A letter was read from the BBKA asking for the nomination of a gentleman or some gentlemen to officiate as judges if required by them and it was proposed that the names of Messrs Scattergood, Pugh and Hayes be given them.

NG 4 Jul 1898
For some days past we have been receiving Press cuttings from all quarters, giving more or less alarming accounts of an incident very regrettable indeed from a beekeeper's point of view, which took place recently near Gedling, a village in Notts. In order to put our readers in possession of both sides of the affair as given, first, by a newspaper reporter and, second, by the owner of the bees which did the mischief, we print below the following from the Notts Daily Guardian of July 4, 1898:
"On Friday, July 1, a very singular incident occurred in a field between the villages of Carlton and Gedling. About 11am. two men named Ablery and Rudkin, in the employ of Mr. Fred Shepherd, timber merchant, were engaged in mowing with a machine, attached to which were two horses, at Bleak Hill. On the hedge-side where operations commenced there is a small spinney and in the spinney an apiary, the owners of which are Messrs. Trimmings and McKinnon.

BEEKEEPING IN VICTORIAN NOTTINGHAMSHIRE

The apiary contains a number of hives, but there is nothing to indicate its existence to the passer-by and the workmen referred to were probably not aware of its existence. As the mowing machine passed along the bees were disturbed. The bees resented the disturbance and fastened upon the horses with great determination and ferocity. In a very short space of time the poor animals were stung in all parts of their bodies and must have suffered agonies of pain. The men attending the machine were also stung, one of them very badly about his head and face.

A man named Elvidge courageously went to the horses and helped to extricate them from the machine, though this could not be done until the arrival of both the owners of the bees. The bees had fastened themselves in the ears of the horses and the wretched animals dashed their heads against the gates and walls, but without avail. They were taken home and Mr. ED Johnson, veterinary surgeon, of Nottingham, sent for.

One of the horses died on Friday evening and the other on Saturday. They were valuable animals and the greatest sympathy is felt with Mr. Shepherd in his loss. The field was not his, but he had engaged to mow it. The man most badly stung was attended by Dr. Knight, of Carlton and is now progressing favorably."

Commenting on the above, Mr. Trimmings who with Mr. McKinnon is referred to as owning the bees in question, writes to correct the reporter's version of what took place. In a note addressed to the editor of the paper in which the above report appeared, he gives the facts of the case as follows:
"About eleven o'clock on Friday morning three men employed by Mr. F. Shepherd, of Carlton, entered a field on Bleak Hill, Gedling and after working twice round the field arrived at a point direct in the line of flight of some bees standing in an adjoining field. It happened that one of the horses was stung by a bee and the men, seeing the other bees passing over, struck at them, consequently they also were stung. Instead of leading their horses away, they themselves ran from the machine and horses to a place of comparative safety and the poor animals were left to the mercy of the bees, which stung them badly, causing them to kick and dance, until they fell, entangled with the harness, etc.

One of them broke away, while the other was helplessly penned down by the pole of the machine for no less than an hour and a half. Then (thanks to Mr. Sketchley, of Netherfield, who rode on his machine and informed me of what had happened). I at once rode off on my bicycle and found one horse being badly stung and the men on the main road, fighting with and terrifying the bees and a crowd all doing the same.

Mr. McKinnon having arrived simultaneously with myself, we at once covered the horse with carbolic and water solution, which had the effect of keeping the bees from killing the animal right out. We then proceeded to cut away the pole and harness and in ten minutes from the time we arrived we had liberated the horse and got it on to the road, quite clear of bees. All this could have been done quite an hour and a half before had we been informed of what was taking place. Had these men taken the horses forward or out of the machine immediately they discovered their danger, instead of fighting with the bees, the serious results and suffering of the poor horses might have been prevented.

The bees were simply working in their ordinary way, and not 'swarming,' as stated, and the swinging of cloths, bags, sticks, handkerchiefs, &c, was the cause of the serious occurrence."

AE Trimmings. Carnarvon Villas, Gedling, Notts.

[Since above was written a circular has reached us notifying that a public subscription is being got up in the neighbourhood where the accident took place to recoup the owner of the horses for his loss. We will be very glad to take charge of any sums which may be sent to this office for the above object, receipt of which will be acknowledged in our columns. Eds.]

BBJ 7 Jul 1898
RM (Newark). We don't quite know what sort of "brass plates for attaching to beehives" you mean, but probably Mr. WP Meadows, of Syston, could supply you.

BBJ 14 Jul 1898
HL (Beeston). Races of Bees. Both samples of bees show a slight tinge of foreign blood, but nothing more than appears in a great majority of the common bees of the country.

NG 16 Jul 1898
The Recent Remarkable Occurrence near Gedling. Sir, The extraordinary attack on two horses by bees at Black-hill, by which the poor animals lost their lives, has naturally been the subject of much comment in the neighbourhood - in fact, it has formed the main subject of conversation. Notwithstanding the explanation given in your columns by one of the owners of the bees, there is a feeling that 25 beehives so near the road and away from any dwelling-houses, constitutes a danger to the public. The field in which the incident occurred has since been mown, but I understand the hives were closed up whilst the mowing and hay gathering took place. This shows the necessity of precautions being taken. But what protection or guarantee is there that if horses or stock were turned into the field and came into contact with the hedge, or made any disturbance, the bees would not again come out in numbers and cause a repetition of the occurrence of July 1st? It was not to be expected that on that day men and horses would go prepared with the apparatus for quelling an attack

of bees; carbolic acid into the bargain. The men naturally buffeted the bees and not being experts, what else could they do? I fail to see how they can be charged with neglect in any shape.

We have been expecting to hear the opinions of beekeepers on this matter, especially as to the likelihood of a recurrence of such an attack. I am not an expert, but I have never heard of an apiary placed in a spinney between two fields and under no control from an adjacent dwelling-house. As a matter of common sense and common prudence, I think the authorities and the owner of the land ought to take the case into consideration.

It is satisfactory to know the bee-owners have signified their intention to assist Mr. Shepherd in his loss and there is no doubt a good subscription will be made. But, certainly, steps should be taken to prevent any possibility of another attack by these useful, but occasionally dangerous, little animals on unoffending horses. Carltonion

BBJ 11 Aug 1898
The sixteenth annual exhibition of bees, honey, and appliances of Leicestershire BKA was held in the show grounds of the Leicestershire Agricultural Society, Victoria Park, Leicester, on July 27 and 28. The entries were not as numerous as in previous years, but the show, on the whole, was very creditable, considering the poor season for honey in the county. Mr. AG Pugh, Beeston, delivered short lectures in the bee-tent and also officiated as a judge.

BBJ 25 Aug 1898
Shropshire BKA, as in former years, held its annual exhibition of bees, honey and appliances, in connection with the splendid show of the Shrewsbury Horticultural Society in the renowned 'Quarry' at Shrewsbury on August 17 and 18. Upwards of 2,000 lbs of honey was staged, but the exhibition lost somewhat by comparison with those held in former years in consequence of the unfavourable season. One of the judges was Mr. P Scattergood of Stapleford.

BBJ 1 Sep 1898
Queens, a few young ones at 3s 6d each. Guaranteed healthy and fertile. A. Simpson, Mansfield-Woodhouse, Notts.

NG 10 Sep 1898
I have received from the secretary of the British Diary Farmers' Association a schedule for their show, which is as usual, to be held at the Agricultural Hall, next month, from the 18th to 21st, and there is an extensive prize-list for honey and products of the hive in various classes and also for displays of appliances. It is to be hoped that some of our Notts. bee men will send up their entries and be successful. The entries

close September 19th.

BBJ 15 Sep 1898
The Derbyshire BKA held their seventeenth annual exhibition of hives, bees, honey and appliances, in connection with the Derbyshire Agricultural Society's Show at Derby, on September 7 and 8. Bearing in mind how adverse has been the past bee-season, it may be said this year's show was quite up to the average. The judge was Mr. P. Scattergood, jun., of Stapleford and he was ably assisted by Mr. R. Giles, of Etwall. Mr. Scattergood also conducted on behalf of the BBKA an examination of candidates for third-class experts' certificates.
Collection of Appliances—2nd, EC Walton, Newark
Six 1-lb Jars Extracted Honey—2nd, W. Lee, Southwell

BBJ 15 Sep 1898
The monthly meeting of the BBKA Council was held at Jermyn-street, SW, on September 9, under the presidency of Mr. TW Cowan (Chairman of the Council). A letter was received from P. Scattergood, Jr expressing regret at enforced absence from the meeting. On behalf of the Education Committee Mr. WH Harris presented a statement compiled from the reports of examiners of candidates for third-class certificates at Nottingham. As the result of this report it was decided to grant certificate of proficiency in the third class to J McKinnon.

BBJ 15 Sep 1898
The NBKA annual show of bees and honey in connection with the Eastwood, Greasley, and Selston United Agricultural and Horticultural Shows was held at Moorgreen on September 6, and was a decided success. The climatic conditions left nothing to be desired and the hundreds of visitors who passed through the tent where the honey was staged were delighted with the exhibits, some of which, considering the season, were very good indeed. Mr. P. Scattergood, junior, of Stapleford, was the judge appointed by the County Association and the following are his awards:
Observatory Hive, with Bees and Queen—1st, A. Warner, Moorgreen;
2nd, G. Marshall, Norwell; 3rd, W. Swann, Eastwood
Six 1-lb Sections—1st, G. Marshall
Six 1-lb Jars Extracted Honey—1st, G. Marshall; 2nd, G. Smith, Bradmore; 3rd, G. Bolton, Eastwood
Six 1-lb Jars Granulated Honey—1st, G. Smith; 2nd, W Lee, Southwell; 3rd, G Marshall
Six 1-lb Jars Extracted Honey—1st, T. Cooper, Lynncroft; 2nd, G. Bolton; 3rd, W. Swann
Frame of Honey—1st, G. Marshall; 2nd, W. Lee; 3rd, W. Swann

BBJ 29 Sep 1898
Young Queens. Two left, 5s each. Guaranteed healthy, fertile, and safe arrival. A. Simpson, Mansfield Woodhouse, Notts.

BEEKEEPING IN VICTORIAN NOTTINGHAMSHIRE

Minute Book – Quarterly Committee Meeting held in the Peoples' Hall, Heathcote-street, Nottingham on Saturday, October 8th, 1898. Present: Messrs Ellis, Forbes, Faulconbridge, Gray, Hardy, Harrison, McKinnon, Marriott, Pugh, Puttergill, Scattergood, Turner, Warner and hon. Secretary. Mr Hardy proposed and Mr Turner seconded that Mr Pugh be asked to take the chair. The minutes of the previous meeting were read when Mr Scattergood proposed and Mr McKinnon seconded, that the chairman be asked to sign the same as correct.

Amongst the correspondence read was a letter from the Sutton-in-Ashfield Society asking for prizes to be arranged for their show. As the application came too late in the season and with only ten days' notice, it could not be done.

Also letters were read about the Derbyshire BKA refusing to accept Mr. J Herrod as a candidate for expert's examination at their show unless he became a member of their association. Mr. Gray proposed and Mr Ellis seconded that the secretary take up the matter with the BBKA and report to the next meeting.

The balance sheet for the quarter was next gone through when Mr. Scattergood proposed and Mr. Harrison seconded that the same be received.

Mr. Scattergood proposed and Mr McKinnon seconded that the secretary be asked to write to the Notts. Agricultural Society about the grant for the shows in 1899 and report to next meeting.
The question of localising the 'Records' was again brought forward and after considerable discussion it was resolved that the secretary make enquiries as to cost and to report to the next meeting.

NG 8 Oct 1898
W. Swann, Eastwood, writes and sends for my inspection a bottle containing two "grubs," which, he says he found between the quilts and frames of one of his hives. He also says the hive and quilts were clean. The grubs are the larvae of the wax moth and you should try to rid your hive of these pests and a ball or two of naphthalene will help you to do this. When these larvae are allowed to have their own way they soon make havoc of the frames and sometimes I have seen them spoil most of the frames in a hive. Several years ago I was at Hucknall listening to JH Howard lecturing in the bee tent and he had been driving a stock of bees from a skep but had failed to find the queen and on carefully examining the skep and taking two combs out which had become detached from the sides of the skep, he discovered a colony of wax moths, which had made their home at the top part of the skep.

To the entymologist these were interesting, but to the beekeeper, as such, very undesirable. In the main these pests are only suffered in weak colonies and they are most troublesome late in the summer or early autumn, when the wax moth may be seen flitting about at the hive door or entrance to gain access to the combs on which to lay her eggs, but if the population be at all numerous, the guards give her but small chance of effecting an entrance: but if, as was the case at Hucknall, the hive, through hopeless queenlessness, has become miserably weak and lost heart, then the moth will obtain an entrance and deposit its eggs on the combs and, if suffered to remain, the larvae worm their way through the midribs constructing a silky tunnel, on the walls of which will he found their dejectments resembling grains of gunpowder and by degrees the comb is utterly ruined. My advice is keep a sharp look-out for these pests, and destroy them when found.

BBJ 20 Oct 1898
The Dairy Show
Interesting Exhibit connected with bee culture. vhc and reserve No., P. Scattergood, Jun., Stapleford.

NG 29 Oct 1898
It will be in the recollection of our readers that a few weeks back I called attention to the Gedling disaster, it may also he remembered that I made an appeal to beekeepers to support a subscription list which had been opened to assist the owner, who lost two horses through the affair. The editors of the BBJ also took up the matter and opened in their journal a subscription list to assist the man and as the whole business has been communicated through the kindness of Mr. George Hayes of the NBKA, I am to present a statement to our readers which I am sure they will regard as eminently satisfactory. A public meeting was held at the Black's Head Hotel, at Carlton on Tuesday evening, October 12th to receive the remaining subscriptions and to wind up the matter Mr JA Hill presiding. Amongst others present were Messrs J Armstrong (hon. Secretary), Wheller (assistant secretary), Osborne (auditor), W. Vickers, C.C. (hon. Treasurer). Messrs Hayes, Pugh and McKinnon (representing NBKA).

From the report given at the meeting it appears that there has been subscribed through the BBJ and from Notts. beekeepers the sum of £26 19s 3d and the amount subscribed by friends in Carlton and district was £26 1s 6d and after deducting expenses a sum of £48 was handed to Mr. Shepherd, the owner of the horses, who expressed thanks and satisfaction for what had been done for him. I think it must be admitted that this is a very satisfactory termination of the affair especially towards the owner of the horses, because it must be admitted that he is no looser by the disaste and further, the owners of the bees were in no way responsible for the disaster, and if the matter had been taken to the courts the probabilities are that the owner of the horses would not have got a farthing because I hold that the whole business was due to the recklessness of the driver of the horses. However, I am glad for the credit of the industry that the subscriptions have come in so well, and that any difficulty that might have arisen has been prevented.

NG 29 Oct 1898
Along with many other beekeepers I visited the Dairy Show last week and was pleased with the exhibits in the honey department and the whole arrangement of the same. There was some splendid honey staged and out of the large number of entries, Notts. was only represented by two entries one of which gained the reserve number and a very highly commended.

BBJ 3 Nov 1898
The monthly meeting of the BBKA Council was held on the 20th inst. at King William-street, Strand, WC. Mr. ED Till occupied the chair and there were also present AG Pugh and P. Scattergood. A letter apologising for non-attendance was received from G. Hayes. The last quarterly Conversazione of the present year was held on the 20th inst. at 4 o'clock, at Jermyn-street, the usual attendance being largely reinforced by visitors from the Dairy Show. Mr. Scattergood said unicomb hives were common in Notts.

NG 12 Nov 1898
The recent attack on horses by bees at Gedling. Fred Shepherd, of Carlton, begs to tender his sincere thanks to the various gentlemen living outside the parish for their kind subscriptions to the fund for re-embursing his loss in connection with the above and he would also especially wishes to acknowledge the good feeling and support given to him by his friends and neighbours of all classes.

NG 19 Nov 1898
A Novice, Carlton, writes: 1st, I am greatly interested in beekeeping and should very much like to keep these most interesting creatures, but do not exactly know when to begin:
2nd, which is the best to begin with, a new swarm, or an established stock?
3rd, what hive should you recommend a beginner to use?
In reply "A Novice"' I am pleased to hear that you are interested in the subject of beekeeping and I am sure you will find it of greater interest the more you know about it. As to the time when you should commence, of course, this must be regulated by circumstances and your own feelings and inclination in the matter, but my advice to beginners is always "Make haste slowly." There are certain reasons why beginners should not purchase their stocks for commencing at the end of the season, because while the bees in winter require only a small amount of attention and care if they have been properly packed up and supplied with food in the late autumn. Still they sometimes do require that attention and care, which in the main only comes from experience and I think the best for beginners is to commence in the spnngtime.

As to which is the best to begin with a new swarm or an established stock, this again is a matter of opinion, but my advice is to purchase a swarm in May, from a hive which swarmed this year and the queen in such a swarm will be at her best. You could then be sure of obtaining good straight clean combs in your hive and it is my opinion that this would pay you much better in the long run than the purchase of an established stock.

As to hive, this is rather a vexed and complex question, because some beekeepers prefer one kind and some another and each believes he has the best. For my own part I prefer the "WBC". Others prefer one less costly, but if you get a reliable makers catalogue you will be able to decide which kind will suit your pocket, but here let me say do not under any pretext whatever buy a cheap and "nasty" hive; purchase a good one and you could give less than 16s for a complete hive. You may have to go higher in price if you have a WBC. It will cost you about 22s and you might order it already and when you get it home give it two or three coats of good paint and let it stand under cover for a few days so as to be quite ready for use when wanted. Then you might get your frames filled with foundation using full sheets in all cases, also be sure to wire all your frames and securely fix your foundation on the same and if you are in any difficulty whatever be sure to write again and I shall be pleased to help you in any way I can.

Minute Book – Committee Meeting held in the Peoples' Hall, Heathcote-street, Nottingham on January 7th, 1899. Present: Mr Scattergood in the chair. Messrs Brooks, Ellis, Faulconbridge, Gray, Hallam, Harrison, Herrod, Jones, McKinnon, Pugh, Puttergill, Rawson, Warner and hon. Secretary.

The minutes of the previous meeting were read, Mr Ellis proposed and Mr Pugh seconded that the same be confirmed. The quarterly balance sheet was read and Mr McKinnon proposed and Mr Herrod seconded the same be received. The annual balance sheet was next discussed when Mr Gray proposed and Mr Turner seconded the same be received and printed with the circular convening the annual meeting.

The question of 'Records' was again brought forward and the secretary gave particulars he had got about the cost of localising and it was proposed by Mr Scattergood and seconded by Mr Ellis that the matter be shelved for the present.

The question of notices offering reward for conviction of persons interfering with or damaging members bees was brought forward when Mr Pugh proposed and Mr Herrod seconded that this meeting do hereby strengthen the secretaries hands and empower him to use these rewards at his discretion.

The arrangements for the annual meeting were next considered and it was agreed

to hold it on Saturday, February 25th at the Victoria Restaurant, Lister-gate at 3pm,
> that Mr Ellis ask the Sheriff (Frederick W Gregory) to take the chair
> that there be at least two competitions – one for liquid and one for granulated honey
> that the prizes be 3/6d and 2/-
> that the honey go to the Children's Hospital
> that Mr Hayes offer of a talk on "Bees and Fruit" be accepted
> that the usual prize drawing take place
> that the secretary provide the little extra prizes needed beyond that generously provided by the president and others
> and that a circular be printed and sent to each member ????? this annual meeting.

NG 14 Jan 1899
Alleged Attempted Destruction of Hives.
A few weeks ago an alleged attempt was made to destroy a valuable apiary at Gedling. A quantity of hay, which had been soaked in tar, was found alight near the hives, but fortunately the discovery was made in time to save the bees. The NBKA have taken the matter in hand and have issued a bill offering a reward for the conviction of the offenders.

Minute Book - Annual Meeting held in the Victoria Restaurant, Lister-gate, Nottingham on Saturday, February 25th, 1899. Mr WS Ellis in the chair.

A letter was read from his Lordship stating that he was on his way abroad and also his services with reference to the annual county show and the affiliation fee. Also one from JR Anderson stating his inability to be present and regret for the same.

The minutes of the previous meeting were read, confirmed and adopted. The annual balance sheet was next gone through and after explanation by the secretary was unanimously received and passed. The report was next read by the secretary. After which it was resolved that the same be printed with the annual report of the association.

The secretary proposed and Mr Pugh seconded that a hearty vote of thanks be accorded Lord St. Vincent for his help and favour during the past and that he be asked to accept the presidency for the coming year. Carried with acclaim.

Mr Hayes was re-elected hon. Secretary and Treasurer, Mr Scattergood as hon. Auditor and Messrs Hayes and Pugh delegates to the BBKA. The committee are re-elected with the exception of Mr Poxon (deceased) and Mr McKinnon who is *ex officio* member and Messrs Lindley and Skelhorn were elected in their place.

His Lordship's remarks about holding the annual show with the Notts. Agricultural Society was next discussed and after deliberation it was decided that we cannot break faith with the NAS seeing we had promised to go with them.

His Lordship's suggestion about the affiliation fee was next well discussed but the meeting was of the opinion that to sever our connection with the parent society would be greatly to our disadvantage.

Mr Marriott brought forward a proposal that the supply of the 'Record' at 1/- per annum to members requiring the same be discontinued as this was a constant drain upon our funds. This was negated and the matter remains as it was previously.

About 75 members and friends sat down to tea at 4.30pm and after tea the meeting resumed when more members came making the total up to quite 80.

The secretary delivered his talk on 'Bees and their relation to clovers, fruit and flowers' illustrated by lantern (slides).

The evening concluded by awarding prizes to winners in competition, distribution of medals won at annual shows and finally by the usual prize drawing. The winners in the competitive classes and drawing being as follows:

1st	Hive	given by the President	won by Mr Swann, Eastwood
2nd	Hive	given by the President	won by Mr Codd, Basford
3rd	Bottle box		won by Mr T Marshall, S-on-T
4th	Hive	given by Mr Varty	won by Mr Bartle, Newark
5th	Honey jars	given by Mr Harrison	won by Mr Mounteney
6th	Smoker	given by Mr Blow	won by Mr Ellis
7th	'Weed' foundation	given by Mr Scattergood	won by Mr Forbes
8th	Guide Book	given by Mr Henton	won by Mr Smeeton
9th	Volume 'Records' 1898		won by Mr Hallam
10th	100 honey labels		won by Mr Henshaw

NG 25 Feb 1899

The annual meeting of the NBKA will be held on February 25th, at the Victoria Restaurant, Lister-gate, Nottingham, at three o'clock. Mr. JR Anderson, JP. is to preside and a very interesting programme, coupling business and pleasure, is arranged. From the balance sheet of the association now before me, I find that the total amount received as subscriptions from members during the year has been £31 4s 6d and from donations to prize funds and other sources £27 4s 4d and the association had a balance in hand at the commencement of the year 1898 of £1 16, making a net income for the year of £60 4s 10d. The expenditure for the year has been £65 0s 11d, leaving a balance due to the treasurer of £4 16s 1d. This is unfortunate and is mainly owing to the fact that the society does not receive that support it deserves. An examination of the balance sheet reveals the fact that the society has only eight members whose subscriptions amount to 10s and over, while there are 53 members who subscribe 5s, and 94 who subscribe only 2s 6d. The society is worked very economically, indeed I am struck in making an analysis of the balance sheet with the amount of work done for the cost. It is to be sincerely hoped that the association will receive the help from those who are in a position to render financial assistance, so that it may carry out its splendid work among the cottagers and artisans of the county during the coming year.

BBJ 2 Mar 1899

The NBKA annual general meeting was held on February 25, at the Victoria Restaurant, Nottingham, Mr. WS Ellis, vice-president, in the chair. Amongst those present were the Rev CWH Griffith, Messrs. Geo. Hayes (secretary), P. Scattergood (auditor), Arthur G Pugh. R. Turner, W Herrod, SW Marriott, MacKender, TN Harrison, GE Puttergill, JT Faulkenbridge, WP Meadows, TJ Waterfield, J Herrod, T Marshall, J McKinnon, AW Codd, Mrs. Pugh, Mrs. Hayes, and Mrs. Harrison.

The annual report, which was read by the Secretary, referred to the season of 1898 as a rather disheartening one for bee-keepers, on account of the exceptionally dark colour of the honey gathered. Mention was also made of the shows held at Colwick, Southwell, and Moorgreen.

The membership for the year ending December last was 156, against 148 of the previous year. Fourteen new members had joined for 1899, making the increase up to twenty-two. The Notts. County Council kindly renewed their grant of £30 for technical instruction in bee-keeping for 1898-9. This had enabled them to have lectures, with demonstrations and the bee tent at Colwick, Welbeck, and Kingston, with lectures at Elston, Lambley, Winthorpe, Costock, Bradmore, North Wheatley, West Bridgford, and Stanton Hill.

The balance-sheet showed total receipts, £65 0s 11d, a balance of £4 16s 1d being due to the treasurer. The report and balance-sheet were unanimously adopted. Viscount St. Vincent, President of the Association, Mr. G. Hayes (secretary), and Mr. P. Scattergood (auditor) were re-elected, Messrs. Pugh and Hayes being appointed delegates to the BBKA.

After the meeting a conversazione was held, and the Secretary delivered an address, which was illustrated by limelight slides. There were also the usual competitions, together with a prize-drawing for handsome and useful presents to members.

NG 4 Mar 1899

A Novice's Questions. A novice, who says he knows nothing about beekeeping, except what he heard at two lectures at the University College a few weeks ago and what he has occasionally read in the "Corner," asks when and how to commence beekeeping. I presume from your query that at present your knowledge of bees and their management is not very extensive. If I am correct. I advise you not to start on too large a scale. Get all the information you can and perhaps the best thing for you to do would be to purchase "The Beekeepers' Guide Book," published by Houlstons, at 1s 6d. If you could get some beemaster or expert to visit your apiary and see for yourself the various manipulations carried out, it would give you assistance and enable you to master some of the details which to the novice appear formidable. I shall be glad in the summer to give by arrangement a day among my bees to those readers who express through the editor a wish to see an apiary in thorough working order and the various manipulations necessary to success. The best time to start beekeeping is at the end of May or early in June with a swarm, whether the hive adopted be a skep or bar-frame one. The apiary being some distance from the house is not a great drawback, except in the height of the season; there is no necessity for daily visits to the apiary. If you will follow the advice given in this column, I have no doubt you will succeed, but do not hesitate to forward queries to the editor whenever you experience difficulty in the management of the bees.

BBJ 9 Mar 1899
The annual general meeting of Leicestershire BKA was held at the Victoria Coffeehouse, Leicester, on March 4. Mr. HM Riley presided. There was a good attendance, amongst those present being AG Pugh (Notts.)

BBJ 4 May 1899
W. Swann (Notts.) Dead Queen Cast Out. The bee sent is a virgin queen. It thus becomes evident that the mother-bee of the colony has met with some mishap, and been replaced by a young queen after all chance of mating had gone by.

BBJ 11 May 1899
The monthly meeting of the BBKA Council was held in the Council Room, Hanover-square, W. on May 5, under the chairmanship of Mr. ED Till. The Education Committee reported they had approved nominations of Judges and Examiners at the County Show in Nottinghamshire.

BBJ 11 May 1899
I was asked the other day by a County Council lecturer on horticulture if I knew any cases where the spraying of fruit trees with such things as "Paris green" or "London purple" had had any injurious effect upon the bees; of course, the lecturer said, " I mean before the trees are in bloom and after the bloom is set, not while the trees are in bloom." I must confess that I have never, to my knowledge, met with a case where

these things have been used and if any of your readers have had any experience in this matter I wish they would give us the benefit of the same in the BBJ; I know that in America this has been a vexed question and that in some States (and the number is increasing) it is made a criminal act to spray fruit trees when in bloom with these poisonous compounds, but what I wish to know is. Have these things been used in England to the injury of the bees in the neighbourhood?

Peter Scattergood, Stapleford, Notts. May 2.

BBJ 18 May 1899
Swarms for Sale, 10s 6d each or 2s 6d lb. Hollingworth, Manor Farm, Wysall, Notts.

Minute Book – Committee Meeting held in the Peoples' Hall, Heathcote-street, Nottingham on Saturday April 1st, 1899. Present: Messrs Brooks, Forbes, Hardy, Hallam, Harrison, Lindley, Marriott, Pugh, Rawson, Scattergood, Skelhorn, Turner, Wilson and hon. Secretary. Mr. Scattergood having been voted to the chair, the minutes of the previous meeting were read and confirmed. The balance sheet was next gone through when it was resolved that the same be accepted.

The sub-committee for dealing with the grant from the Notts. County Council was next dealt with when it was resolved that the existing committee should be re-appointed 'en bloc'.

Correspondence was next read and amongst the same was a letter from Mr. Warner of Moorgreen who stated he was leaving the county. Mr. Scattergood proposed and Mr. Pugh seconded and Messrs Forbes and Turner supported a resolution that the secretary be asked to write to Mr. Warner expressing the regret of this meeting on learning of his removal from amongst us.

A discussion here followed with regard to mode and amount of grant to each show and it was finally agreed to take each show on its own merits and that no hard and fast line be laid down.

An application from the Beeston Society was next considered and it was resolved that we accept and pay one fourth of the prize money awarded. The Beeston Society to have the entry fees. Resolved that we join the Moorgreen and Southwell Societies on the usual terms.

The schedule for the annual show at Mansfield was next gone into, and altered to suit the present requirements and ordered to be printed as soon as all arrangements were complete.

Resolved that the judges for the shows be as follows:

Mansfield	Either Mr. Cribb, Dr Sharp or Mr White in the order named.
Beeston	Mr. Scattergood
Moorgreen	Mr. Turner
Southwell	Mr. Scattergood

The secretary stated that it was possible that the Clarboro' Society would ask us to join them and it was decided to leave the matter in the hands of the secretary.

BBJ 2 Jun 1898
DC (Mansfield) The sample sent was affected with foul brood.

NG 19 Jun 1899
On Wednesday the Nottinghamshire Agricultural Society celebrated its majority by the holding of its 21st annual two days' exhibition upon a well-arranged ground at Mansfield and, in welcome contrast to the experience of some recent years, the weather was gloriously fine.

As usual a good deal of interest was manifested in the competition arranged under the auspices of the NBKA. Altogether there were thirteen classes, but the entries hardly compared favourably with those in preceding years. It is to be hoped that this is not attributable to any falling off in the popularity of beekeeping. The prize winners were limited to the names of a few prominently associated with beekeeping in Nottinghamshire who have been successful upon previous occasions. In response to the substantial prizes offered by the association some excellent exhibits were staged, particularly noticeable amongst them being in the attractive display by Mr. JT Faulconbridge, of Bulwell, and an admirable collection of hives and appliances, including several recent improvements and developments. At regular intervals Dr. Sharp, of Brant Broughton, Lincolnshire delivered lectures upon the subject of beekeeping, the discourse being accompanied by a practical demonstration. A good number of visitors interested in the art profited considerably from the instruction thus conveyed. That gentleman also officiated as judge, the stewards being Mr. Hardy (Sutton), Mr. McKinnon (Gedling), and Mr. Maskery (Sutton).

NG 17 Jun 1899
Honey on Sale.
 WBM. Kirklington says he has a few stones of extracted honey for sale. It is a bit brown, but good, and he wants to know if I can tell him of a market where he could sell it?
I cannot now tell you of a market for your honey, but if you had written four months back I might have helped you; moreover, you do not say what price you want for it. You might place an advertisement, say for a week, in our daily issue.

BEEKEEPING IN VICTORIAN NOTTINGHAMSHIRE

NG 24 Jun 1899

It was a great disappointment to me that I could not get to Mansfield to the county show. It is the first time for many years that I have missed the annual show of the NBKA, but the indefatigable secretary, Mr. Hayes, has himself sent me a few lines respecting the same, which I gladly give for the benefit of our readers who, like myself, could not get to the show. He says:

"The entries were not quite so numerous as usual, owing to the last season being so bad, but some educational and interesting exhibits were lent by Messrs G Hayes, P. Scattergood and W. Herrod, and these helped to make the show interesting to the many visitors to the tent in which the exhibits were staged. The bee tent was as usual a centre of attraction and Dr. Sharpe and Mr. FJ Cribb, both neighbours and members of the Lincolnshire BKA, lectured and gave demonstrations to the large audience that gathered round the netted enclosure. Dr. Sharpe officiated as judge of the exhibits."

I am also reminded that there are several other shows in the county, at which lectures and demonstrations are to be given, viz. Beeston, Welbeck and others but more of this nearer the time.

NG 1 Jul 1899

JB, Trowell, has a stock in a hollow tree and he wants to get them out and wishes to know what he is to do:

Swarms that abscond will frequently go in an almost straight line towards a new home, be it in a tree or hollow in a wall or elsewhere and the inference one draws is that they are being led to a spot previously selected by scouts. This is evidently the case and we may often see bees about some hole in a tree which probably contains combs and on watching them we shall find that they are "cleaning up" preparatory to a swarm taking possession of the hollow. I have seen swarms go in this manner after their owners have felt sure they were comfortably settled at home. And there they must stay if they choose an awkward spot, as it is a most difficult thing to get them out after once they get settled inside. Suppose we follow a swarm and arrive almost as soon as the bees get inside the hollow of a tree. We might force them out by smoke or the fumes of carbolic add, particularly if there is a hole above the one by which they entered. When they have settled for at least a season their removal becomes almost impossible, unless the tree is cut down and broken to pieces. As it is only the bees "JB" wants, his best plan will be to fix a skep or box in front of the hole in such a manner that the bees are compelled to pass through in order to get to the open air. When they become crowded inside the tree and would almost to a certainty swarm, they all spread out and occupy the box or skep in front of their crowded home. Their combs will be built and ultimately that temporary structure will become the brood chamber. After it has been occupied for some time by bees and queen, it might be removed, or left there until the end of the season and carried away with the bees comfortably settled inside.

BBJ 6 Jul 1899

The NBKA annual Show was held in conjunction with the Nottingham Agricultural Society at Mansfield, on June 7 and 8. The entries were not quite so numerous as usual, owing to the last season being so bad, but some educational and interesting exhibits were lent by Messrs. Hayes, Scattergood and Herrod and these helped to make the show interesting to the crowds who visited the tent. Dr. Percy Sharpe and Mr. FJ Cribb, gave demonstrations in handling bees and lectured to large audiences in the bee-tent each day. Dr. Sharpe also officiated as judge and made the following awards:

Honey Trophy—1st, JT Faulconbridge, Bulwell

Twelve 1-lb Jars Extracted Honey (light) —1st, J. Herrod, Sutton-on-Trent; 2nd, JT Faulconbridge; 3rd, P. Scattergood, Stapleford

Twelve 1-lb Jars Extracted Honey (dark) — 1st, J. Herrod; 2nd, JT Faulconbridge; 3rd, GE Puttergill, Beeston

Twelve 1-lb Sections —1st, H Merryweather, Southwell; 2nd, GE Puttergill

Twelve 1-lb. Jars Granulated Honey — 1st, J. Herrod; 2nd, Geo. Smith, Bradmore; 3rd, H. Merryweather

Shallow Frame of Honey —1st, W. Lee, Southwell

Honey Vinegar —1st, P. Scattergood.

Beeswax —1st, No award; 2nd, Geo. Marshall, Norwell

Observatory Hive —1st, G. Marshall; 2nd, J. Herrod

BBJ 6 Jul 1899

The seventeenth annual exhibition of bees and honey of Leicestershire BKA was held in the Show Ground of the Leicestershire Agricultural Society, Leicester, on June 23 and 29. The exhibition was favoured with congenial weather on both days and a good attendance resulted. The manipulating tent came in for a fair share of patronage on both days. Lectures were delivered at intervals by Mr. Peter Scattergood, Stapleford who was assisted by the hon. secretary, Mr. J Waterfield, as manipulator. Mr. Scattergood officiated as judge.

NG 22 Jul 1899

I have recently paid a visit to a number of beekeepers in a distant part of the county and could not help remarking the striking difference in some of the apiaries. One beekeeper I visited had about six or eight hives, all trim, neat, and tidy, nicely painted; clean inside and out and the quilts and coverings all in good order and made to look as nice as one could wish, while at another apiary not a mile away, I found dirty hives full of cobwebs inside, between the hive and outer casing, with quilts that looked as though they had been in existence for ten years and packing of a fearful and wonderful character, from a child's pinafore to a dirty, old, ragged coat. The hives presented a neglected appearance and were full of cracks and looked as though they would tumble to pieces; in fact, the whole apiary presented a most neglected and miserable appearance and I was told that the bees had not done well for several years and no wonder. To me it was a marvel that the bees had lived through it all. Beekeeping will not pay under such conditions and I am also equally sure that it will pay to keep the apiary clean and tidy.

I also found on my tour that honey of a good quality is coming in rapidly and that there is every prospect of a fairly good harvest. A correspondent from Newark writes me this week that the new honey is good both in colour and quality in that district and as I intend to visit the show at Southwell on Thursday next, I hope to be able to give my readers some idea of the quality of the honey in that part of the county, which, among bee men is considered the best for honey production. I am also informed that there is to be a show of honey and bees and demonstration at Beeston on August 5th, and I shall endeavour to be present and hope to meet a good number of the craft.

BBJ 27 Jul 1899
The NBKA annual show of bees and honey in connection with the Southwell Horticultural and Cottage Gardening Society was held on July 20, and may be regarded as most successful. Mr. P. Scattergood officiated as judge and made the following awards:
Shallow Frame of Comb Honey—1st, GH Pepper, Winkburn Hall; 2nd, H. Merryweather, Southwell; 3rd, W. Lee, Southwell
Six 1-lb Sections—1st, H. Merryweather; 2nd, JT Faulconbridge, Bulwell Wood
Six 1-lb Jars Granulated Honey—1st, J. Herrod, Sutton-on-Trent; 2nd, JT Faulconbridge; 3rd, G. Smith, Bradmore
Six 1-lb Jars Extracted Honey—1st, J. Herrod; 2nd, JT Faulconbridge; 3rd, Geo. Bell, Cottam; 4th, W. Lee
Observatory Hive, with Bees and Queen — 1st, Hy. Merryweather; 2nd, Geo. Marshall, Norwell
Six 1-lb Jars Extracted Honey (non-winners at previous shows only)—1st, J. Breward, Staythorpe; 2nd, GH Pepper

BBJ 27 Jul 1899
Wanted, Strong Swarm (Ligurians preferred) in exchange for British Bee Journal, 1885-6-7-8, to July, '89, pure '99 Queens. Hill, The Park, Kirkby.

BBJ 17 Aug 1899
Under the auspices of the Lancashire County Council, a show of honey hives, etc., with lectures and demonstrations of beekeeping, was held at the above Society's exhibition in the Wavertree Recreation Ground, Liverpool. Lectures and demonstrations were given at frequent intervals each day by Mr. Herrod and were attended by large numbers, the fine weather being most suitable for the manipulations.

Articles of Food, etc. in which Honey is an Ingredient—1st, Peter Scattergood, Stapleford, Notts.

BBJ 17 Aug 1899
Glorious weather favoured the annual flower show and gala held on August 8 and 9, at the Abbey Park, under the auspices of the Leicester Corporation Parks Committee. An exhibition of bees, honey and appliances was held in a special tent, under the management of the Leicestershire BKA. The lectures and demonstrations in the manipulating tent by AG Pugh, Beeston, attracted many visitors and created much interest. The entries exceeded those of last year and the quality of the honey was quite up to the average. Mr. Pugh kindly officiated as judge.
Cake, not less than 1 lb sweetened with Honey—1st, Peter Scattergood, Stapleford

NEP 17 Aug 1899
Welbeck Tenants Show. It is disappointing to see what slight evidence there is that beekeeping flourishes in the Midlands, there being, in point of fact, only eight entries by the tenants and very few by the cottagers. Yet one would think that old Sherwood Forest, with its mingling of sylvan arable and pastoral, would afford fruitful pastures for the bees, which never introduced anywhere without benefit the orchard and fruit gardens.

NG 19 Aug 1899
A fortnight ago I referred to this subject and I wish now to say that so far as I can gather from the many beekeepers I meet from time to time and from those of my readers who now and then send me a letter I gather that in this district, ie. in the county of Notts. the season of '99 has been a fairly good one; while for quality it is somewhat above the average. The limes in my immediate neighbourhood have bloomed splendidly and the bees have worked on them with a right good will and have carried large quantities of nectar into the hives. We also have had a moderate yield from the clover, though the hot weather has almost burnt it up in many places. Beekeepers will be soon winding up their season's harvest and will be striking a balance of profit and loss. I shall be glad to hear from any of my readers as to the quantity and quality of their honey.

NG 2 Sep 1899
Under very favourable auspices, the fourth annual Aspley and District flower, fruit, and vegetable exhibition was opened in a large field near the Wheat Sheaf Inn, Bobber's Mill, there being a large attendance during the afternoon and evening. During the afternoon a lecture on apiculture was given by Mr. G Hayes, secretary of the NBKA who demonstrated "How bees could and should be kept."

BBJ 7 Sep 1899

ECS (Notts.).—1. Comb contains no brood at all nor any trace of such. Although old and needing renewal there is nothing worse in cells than honey and pollen.
2. We cannot detect any "putrid kind of smell" in comb as stated.

NG 9 Sep 1899
Greasley, Selston and Eastwood have held a show under the auspices of a well-managed organisation, which embraces the three parishes within the scope of its operations and in exceedingly favourable, if somewhat oppressive, weather, the 49th exhibition was held at Moorgreen on Tuesday. An attractive feature was found in a series of competitions organised in connection with the NBKA. The show of bees and honey was considerably above the average, both in quantity and quality.

NG 16 Sep 1899
I received by post the other day the schedule of the British Dairy Farmers' Association Show, to be held at the Agricultural Hall, Islington, from 17th to 20th October and quite a number of prizes are offered for honey and honey products. The time of closing for entries is Monday next, September 18th. I hope some of our Notts. showmen will be to the front this year.

BBJ 21 Sep 1899
The eighteenth annual show of hives, bees and honey of Derbys. BKA was held in connection with that of the Derbyshire Agricultural Society at the Cattle Market, Derby, on September 13 and 14. Mr. Peter Scattergood, Stapleford, Notts, was the appointed judge.

BBJ 28 Sep 1899
HCW (Notts.). Queen Found Dead in Introducing Cage.
1. The queen has the appearance of having been starved to death. It was unfortunate not to have made sure the stock was not queenless before trying to introduce the stranger,
2. Mr. Geo. Hayes, Hon. Sec. NBKA, 14, Mona-street, Beeston, will no doubt supply the information needed if written to.

BBJ 5 Oct 1899
Under the auspices of the Lancashire BKA, one of the most successful and interesting exhibitions of honey and honey products was opened at the St. James' Hall, Manchester, on Monday, September 21.
Honey Vinegar — 1st and 3rd, P. Scattergood, Stapleford, Notts.

Minute Book – Quarterly Committee Meeting held in the Peoples' Hall, Saturday, October 7th, 1899. Present: Messrs Brooks, Forbes, Faulconbridge, Gray, Harrison, Lindley, McKinnon, Mackender, Pugh, Puttergill, Rawson, Skelhorn, Scattergood,

Turner, Wilson and Herrod.

Mr. Pugh having been voted to the chair, the minutes of the previous meeting were read. Mr. Scattergood proposed and Mr. Lindley seconded that the same be confirmed.

Mr. Scattergood proposed and Mr Skelhorne seconded that the correspondence be dealt with after the other business of the meeting. Mr. McKinnon proposed and Mr Harrison seconded, an amendment that it be taken in its usual course as set forth in the agenda. The proposition was carried.

The quarterly balance sheet was next gone through when Mr. Scattergood proposed and Mr. McKinnon seconded that the same be received and adopted.

A letter was next read from his Lordship, Viscount St. Vincent, stating his desire that we should find a fresh president. Mr. Scattergood proposed and Mr McKinnon seconded that the secretary write to his Lordship and express the regret of this committee at his intention and also that they earnestly beg that he will reconsider it and retain the office. Mr. Harrison proposed an amendment that a deputation wait upon his Lordship and convey to him personally the wishes of this committee. The original proposition was carried with the addition that the secretary should state that the cheque which had appeared in the current account had opened their eyes to the lavish liberality of his lordship and that they felt they had no right to trespass on that liberality to the extent they were doing.

The 'Bath Bee Case' was next considered and the committee felt that it would be best not to probe into this question; at the same time regretting that it was a case which would not help much to the upbuilding of beekeeping.

The question of correspondence arising out of the minutes of the previous meeting. After hearing both sides of the question and with the consent of both parties, it was decided to burn all the correspondence and let the matter drop.

BBJ 26 Oct 1899
The twenty-fourth annual show of the British Dairy Farmers' Association opened at the Royal Agricultural Hall, London, on October 17, and was continued till the following Friday.
Twelve 1-lb Jars (Light) Extracted Honey -   hc. AG Pugh, Beeston; P. Scattergood, Stapleford
Interesting and Instructive Exhibit of a Practical Nature - 1st; P. Scattergood

BBJ 26 Oct 1899

BEEKEEPING IN VICTORIAN NOTTINGHAMSHIRE

The monthly meeting of the BBKA Council was held on the 19th inst. at Jermyn-street, St. James', the Hon. and Rev. Henry Bligh in the chair. There were also present AG Pugh and Peter Scattergood. At 5 pm. the members and visitors reassembled for the conversazione and among those present were George Hayes, AG Pugh and Peter Scattergood.

NG 28 Oct 1899
Along with several Notts. beemen I visited the Dairy Show last week and was much pleased with the exhibits staged in the honey department. This department is admitted by all I have heard give an opinion on the subject to be a very useful adjunct to this interesting show and perhaps at no other show in the country do we get such an interesting and representative display of honey and honey products. The exhibits were large and good and the Notts. beekeepers were represented by four entries but only three were staged, one of which, an instructive example of a practical nature connected with bee culture, took the first prize. The other two exhibits in the class for 12 bottles of light honey, and which had 52 entries staged, were awarded a highly commend and a commended card. The quality throughout was of a high order.

BBJ 16 Nov 1899
Judging Honey. Mr. Hammond in his letter (p. 434) makes some very strong observations upon Mr. Hooker's statements to the effect that such admissions as were made by the latter gentleman are really placing "a premium upon dishonesty," and asks, "Why," among so many gentlemen of high integrity and standing as were at the conversazione, "such a statement was allowed to pass unchallenged and unrefuted?"

A meeting lasting three hours has necessarily to be very briefly reported in such a small paper as the BBJ, but Mr. Hammond will—if he reads the report again—see that I immediately raised a protest against Mr. Hooker's statement and explained my usual course of procedure when judging honey in sections (as stated on page 422), My practice is to select those which are likely to figure in the prize list and to sample the contents before finally allocating the prizes and I further illustrated the necessity for this by giving an instance where the prize sections at one of our most important shows some years ago were practically worthless, the honey having been contaminated by some foreign substance and the fact being overlooked through the judges not sampling the honey.

The argument that sections should not be damaged by cutting one of the dozen is a poor one because so little damage need be done and, as Mr. Gordon remarks, the prize-cards should surely be of some service in directing purchasers to the best article, I am of opinion that the applause that greeted my remarks proved that my methods were acceptable to the great majority present and I think Mr. Hammond need not fear the dishonest practices he suggests, as it will only be very poor judges who do their work in the slip-shod manner mentioned.

On another point I think if some suitable method is adopted to show the dividing line between light and dark honey classes and the judges are instructed that colour is not to be the sole thing looked for - as too often appears to be the case - it seems to me that the schedule need not be burdened with another class for extracted honey; but I am certainly of opinion that more prizes, or prizes of greater value, should be given. For instance, in Class 69 this year the entry fees at the Dairy Show came to exactly twice the amount of prize-money given and when one considers the great expense incurred in getting exhibits from such long distances, it seems questionable whether entry fees are not too high. One shilling entry for each exhibit would nearly have paid prize-money in some classes this year at the show in question to say nothing of the larger number of exhibits that would be forthcoming with a smaller entry-fee. A better position for the honey and attention to these matters will, I hope, enable the Dairy Show to maintain its reputation as the best show of "bee produce" in the kingdom.
AG Pugh, Beech House, Beeston. November 10

BBJ 4 Jan 1900
"I have been sending flowers and fruit to a firm in St. John's Market, Liverpool, throughout the year and finding them reliable and honest in their dealings I asked them if they could sell honey for me. They replied that they thought they could. On October 27 I sent them some 1-lb and ½-lb pots. They realised 8s per dozen for the 1-lb ones and the same rate for ½-lb. I have sent two consignments since. I have also been sending flowers, vegetables and fruit to a tradesman in the same line in Market-street, Nottingham and he has sold several consignments of extracted honey at 8s per dozen 1-lb jars. This may not be a high price when carriage and commission is deducted, but as I pointed out before, you await no orders, but send when you want to get rid of the honey or need the cash. Of course no one can guarantee the price, but the Liverpool firm have since written to me saying they were taking another bee-keeper's honey from Wales, whom I had recommended to them, and my Nottingham customer has written asking me to send more honey, which I cannot do, as I have not much left." WJ Belderson, Terrington, Norfolk, December 25, 1899.

Minute Book – Quarterly Committee Meeting held in the Peoples' Hall, Heathcote-street, Nottingham on Saturday, January 7th, 1900. Present: Messrs Ellis, Forbes, Faulconbridge, Hallam, J and W Herrod, Pugh, Puttergill, Timmings, Wilson, Rawson, Skelhorn, Scattergood and Wilson. Mr Scattergood was voted to the chair.

The minutes of the previous meeting were read, confirmed and signed. At this point Mr Ellis came in to the meeting and Mr Scattergood vacated the chair in his favour.

The quarterly balance sheet was next read and Mr Pugh proposed and Mr Harrison seconded the same be adopted.

The secretary then went through the annual balance sheet and after going into the various items Mr Scattergood proposed and Mr Pugh seconded that the same be received as satisfactory and that it be printed on the circular convening the annual meeting to members.

His Lordship's reply about the presidency was next read and it was ordered to the annual meeting as per request.

The arrangements for the annual meeting were next considered and it was ordered that the meeting be held on Saturday, February 2nd at 3pm in the Peoples' Hall
> that his Lordship be asked whether it will be convenient for him to be present and if he cannot that the secretary write to the Mayor of Nottingham (Frederick R Radford), the Sheriff (James Brown Sim), Lord Belper or Mr Jesse Hind to preside
>
> that the secretary make arrangements with some caterer to provide tea
>
> that two prizes be given for a single jar of granulated honey and that the honey be given to the Children's Hospital, Nottingham. 1st prize – medal; 2nd prize pendant; or 3/6d or 2/6d
>
> that prizes of 7/6d; 5/-; and 2/6d be given for the best original essay on 'How to become a successful beekeeper'; the essays to become the property of the association - open to members of this association only (to compete under a *nom-de-plume*). Winner to be at the annual meeting to claim and read his paper (or by proxy) and not to occupy more than 15 minutes

The prize drawing was next considered and it was ordered that the secretary obtain a hive value 15/- if necessary and put in some labels for another prize. Mr Harrison promised a table honey jar value 6/6d, Mr Scattergood something value 5/-, Mr Trimmings something value 10/-.

The meeting then terminated with a vote of thanks to Mr Ellis for presiding.

BBJ 18 Jan 1900
"Learner and Starter" (Notts.)
1. It is very unwise for a learner—or, indeed, any one—to do any such examining of bees as will require removal of sections, because in hives on which surplus-chambers are being filled should not be interfered with so far as regards meddling with or upsetting brood chamber. Such undue interference is one of the strongest incentives to robbing.
2. No artificial feeding should be carried on while stocks are storing in surplus-chambers.

3. Bee-candy is given in autumn merely to supplement stores suspected to be insufficient food; the amount of candy therefore depends on the weight necessary to make up a full supply.

Minute Book – Annual Meeting held in the Peoples' Hall, Heathcote-street, Nottingham on Saturday, February 10th, 1900, the Rev HL Williams presiding.

The correspondence which had passed between the president and committee about a fresh president was read, when it was decided to consider when electing officers.

The minutes of the previous annual meeting were read. Mr Scattergood proposed and Mr Pugh seconded that the same be confirmed by the chairman. The annual balance sheet was next gone through and the same being of a satisfactory nature, Mr Scattergood proposed and Mr W Herrod seconded that the same be received and adopted.

At this point Mr Scattergood noticed Mr FHK Fisher, a life member and late secretary, in the meeting and rose to propose "that Mr Fisher be asked to take a post of honour by the chairman." Mr Rawson seconded. After Mr Fisher had taken his place the secretary read his annual report and Mr J Herrod proposed and Mr Fisher seconded the same be received and ordered to be printed with the balance sheet.

The election of officers was next proceeded with and it was proposed that a vote of thanks be accorded the Right Honourable Lord St. Vincent for his past services as president of this association and for his generous help and that he be elected president for the ensuing year. Messrs Hayes, Gray and the chairman supported the resolution which was carried with cheering.

Mr Hayes, proposed by Mr Scattergood and seconded by Mr Mackender and supported by Mr Turner was re-elected secretary and treasurer. Mr Scattergood was re-elected auditor and Messrs Hayes and Pugh, delegates to the BBKA meetings. It was proposed – on each individually - that the following gentlemen form the committee for the ensuing year:
Messrs Faulconbridge, Wadsworth, Forbes, Harrison, Marriott, Lindley, Skelhorne, Pugh, Turner, Puttergill, Wilson, White and Swann.

The question of next year's annual show was discussed when Mr Scattergood proposed and Mr Marriott seconded that if the Notts. Agricultural Society offer us the same terms as previously, we join them with our annual shows and the committee be instructed to arrange the details.

Considerable discussion followed as to the awarding of the BBKA silver medal. A proposition was ultimately put forward that the winners of the silver medal be debarred from it for two consecutive years with a view to enlarge the sphere of possibility for other members. An amendment to this was proposed limiting the time to one year and for which only 5 voted. The original proposal was carried.

The meeting then adjourned for tea at which 62 members and friends sat down.

At 6pm the meeting was resumed and after Mr Ellis had occupied the chair, he distributed medals as follows:
    BBKA Silver Medal to Mr J Herrod, Sutton-on-Trent
    BBKA Bronze Medal to Mr J Herrod
    BBKA Certificate to Mr J Herrod
    NBKA Silver Pendant for Trophy to JT Faulconbridge, Bulwell

The awards in the competitions were next made known as follows:
Class for single jar of granulated honey (for Children's' Hospital) (13 entries)
    1st   BBKA Bronze medal to Mr Scattergood
    2nd   BBKA Silver Medal Mrs Scattergood
For essay on 'How to become a successful beekeeper)
    1st   7/6d to Mr FHK Fisher, Ewerby
    2nd   5/- to Mr Hesslewood, Dunkirk

The papers were next read by the writers and evoked a good deal of discussion which lasted until after 8pm. The meeting then terminated by the usual prize drawing which resulted as follows:

1st Meadows non-swarming WBC hive	given by Lord St. Vincent	won by Mr King, Retford
2nd Meadows non-swarming XL all hive	given by Lord St. Vincent	won by Mr Rodgers, Kelham
3rd Varty's 10/6d hive	given by Mr Varty	won by Mrs Hallam, Orston
4th Parcel useful articles	given by Mr Trimmings	won by Mr Swindon, Clarboro'
5th Fancy Table honey jar	given by Mr Harrison	won by Mr Lacey, Retford
6th Parcel of 'Weed' foundation	given by Mr Scattergood	won by Mr W Herrod, S-o-T
7th 200 Honey labels		won by Mr Hornelow, Mansfield
8th 100 Honey labels		won by Mr Henson, Strelley

    117 members paid and took part in the above drawing

NEP 10 Feb 1900
The annual meeting of NBKA was held at the People's Hall, Heathcote Street this afternoon. The Rev. HL Williams presided over a large attendance and amongst those present were Messrs. WS Ellis, G. Hayes, P. Scattergood, FHK Fisher, Herrod, Harrison, R. Mackender, S. White, R. Turner, SW Marriott, and J. Gray. The secretary (Mr. Hayes) presented the annual report, which recorded a steady growth in the association. The number of members amounted to 132, as opposed 147 three years

ago. The financial statement showed small, but satisfactory balance. The Notts. County Council continued their grant of for technical instruction in beekeeping and with its aid they had been enabled to give practical lessons in the bee tent at Mansfield, Kingston and Clarborough. The Nottingham City Council had also made a grant of £2 2s and arranged for the delivery of two lectures at University College.

The election of officers for the ensuing year resulted as follows: President, Lord St. Vincent; secretary and treasurer, Mr. G. Hayes; auditor, Mr. P. Scattergood; delegates to the BBKA, Messrs. Hayes and AG Pugh; committee, Messrs. Brooks, Faulconbridge, Wadsworth, Forbes, Harrison. SW Marriott, G. Skelhorne, AG Pugh, JR Turner, GE Puttergill, W Wilson, S. White, and W. Swann. The silver and gold medals and certificate awarded by the BBKA for honey were won by Mr. J Herrod, of Sutton-on-Trent.

Articles were contributed for the annual prize-drawing by Lord St. Vincent and others.

NG 17 Feb 1900
How to Become a Successful Beekeeper
The following interesting paper was read at the usual meeting of NBKA on Saturday last and was one among several papers sent in for competition on the above subject. It was adjudged to be the best sent in and obtained the first prize. It is written by Mr. Frank HK Fisher, a former secretary of the Association, late of Farnsfield, but now of Ewerby. Lincolnshire. In compliance with the wish of a number of beekeepers. I print the same in the corner and next week I hope to print the second prize paper:

What is a successful beekeeper? One who manages to get a good return for his outlay with the maximum of honey with the minimum of expense and stings. How to become such I propose to take under ten heads.

1. The Man — Anyone can keep bees, but only a small percentage of those who do can honestly be called beekeepers. A nervous, irritable man cannot become a successful beekeeper, for his manner would upset the bees that they would be unmanageable and not get the necessary attention. The beekeeper should be of a calm disposition — one who loves his bees strives to learn all he can about them, looks after them when they require it and abstains from disturbing them when not necessary.
2. Locality — Although a good beekeeper may do badly in a poor locality, yet to do well the locality must be a good one. Give me a district where are some good orchards, pastures, rich in white clover, an avenue of lime trees, and a common covered with heather. In case you should think I do not require enough, I would add a field of sainfoin, same aconite under the trees, the latter for early pollen, a field of beans and another of mustard. These combined would make an almost ideal bee locality. With such an assortment every season would give

a profitable return. For, if the weather was unsuitable during the flowering of some, it would hardly be during all. If a man be starting beekeeping for a living it is absolutely necessary in a climate such as ours, that he chooses something like I have mentioned, but the majority of us keep bees as a hobby and an addition to our income and not for a livelihood, so we are obliged to make the best we can of the locality we reside in. When I lived in Notts. the district was good for quality and quantity where I am at present living is only moderate in both.

3. The Hive — To an association of beekeepers it almost goes without saying that bar frame hives must be used and expense of outlay is an important consideration if our beekeeping is to pay. The competition amongst manufacturers of bee appliances is so keen that hives are turned out with but a small margin of profit to the makers and an amateur carpenter saves very little (nothing if he reckons his time) if he buys new wood and makes his own. At least that is my experience. Old packing cases broken up give unsatisfactory results often when made into hives, but I have lately found out that the boxes in which Quaker Oats are packed are easily converted into serviceable and good hives. Three boxes will find about all the wood required for making two hives. These boxes are just the right width to take a standard frame and although being too deep, being dovetailed, they are easily knocked into pieces and the sides cut down. On putting them together I find they require nailing. Every super, of course, should use standard frames. Unless he does so he will find that the various etceteras he requires will cost him more, as dealers keep those in stock to fit the standard size. To a beginner I recommend buying the inexpensive hive from some well-known maker and noting with his Guide Book (Cowan's) the little points as to size and spaces. Then, if he feels inclined he can start to make his own. Most hives are made to contain ten frames, but those made from Quaker Oats boxes hold seventeen and a dummy board. With the latter the hive can be contracted to hold any less number. So long as the internal measurements of the hive are correct and the hive itself is dry, bees will do well in it. I have met with some big takes from very rough hives, but never where the hives got wet inside.

4. The Extractor - A honey extractor is a necessary and here lowness of price should not be considered, one with gearing being preferable.

5. Foundation - The use of comb foundation is conducive to success. It is no economy to be stingy in its use.

6. Manipulating - While giving his bees all necessary attention, the beekeeper must avoid undue meddling with the bees, just because he finds manipulating interesting. A stock which is doing well should have its brood nest examined in spring and autumn, once at each time and at other times should be left undisturbed. Of course; if the bee-keeper goes in for queen rearing or introducing new queens, the brood nest would have to be disturbed, but in

ordinary way, twice a year is sufficient.

7. Queens - A very important item towards success to have nothing but young queens. A hive with a young queen nearly always does well, by way of an example, see how a cast of the previous year makes headway in the spring. Also, let a bee-keeper notice those hives which give little produce in the season and he will generally find that the queens are the oldest in his apiary. Every bee-keeper should know the age of all his queens. It is very useful to have some form of memoranda, for each hive, either inside the hive itself or in a book kept for the purpose. Particulars of queen's age, should be there noted. Stocks headed by young queens are not so liable to take diseases as others. And this is an important consideration.

8. Food - Then as regards feeding I am not an advocate for taking all the honey that can be got and then feeding up the bees with syrup, for I consider it takes too much out of the bees and harms the stock, but feeding is one of the important points in successful beekeeping. Some stocks during the winter consume more than others and when the bee-keeper makes his spring examination he should see these and, if necessary, commence to feed with syrup not in large quantities, for at that season of the year we do not wish the bees to store the honey only to keep them going until they can gather honey for themselves. But, whether short of food or not, all stocks should have a little food, the bee-keeper commencing to feed about a month before he expects his main honey flow. This stimulates the queen to lay and so the stocks are strong to take advantage of the honey flow. If this feeding is neglected, many stocks have to do the building up whilst the honey is coming in and in consequence the beekeeper's return is not so great. If the honey flow is short, as may be, very small indeed. Oftentimes, also in the spring, when a stock is in full breeding, a few days' bad weather may come and if the bees are not fed all chance of honey from that stock may be gone, or even the stock. Then with regard to autumn feeding - soon after removal of the supers very little honey is in the body of the hive, the queen having monopolised those combs for breeding and the bees having stored the honey in the supers. As soon as he perceives this, the bee-keeper should feed as quickly as possible until the bees have sufficient for the winter. I say at once and quickly, the sooner it is done the less chance there is of robbing and the bees are able to seal up the syrup.

9. Honey - Having got the honey, the question arises of how to get rid of it and this may be the point between beekeeping paying or being a loss. Nothing is gained by rushing the honey on to the market as soon as obtained, but sections should not be kept long. Prices are generally low while honey is plentiful and it is nearly always so at the end of the season. Sections should always be carefully cleaned of all propolis and dirt and either sold or put in cases before selling. The extracted, if bottled, should be neatly done and the bottles well cleaned: if tied over with parchment, this should be carefully trimmed. The

finish should be given by putting on the County Association label. I will give an instance, a very extreme one, perfectly true, to prove that "get-up" counts. About 12 years ago a Honey Fair was held in Nottingham to help members of the NBKA to dispose of their honey. The highest price for a single 1 lb bottle was 2s, the lowest $3^1/_2$d, while the honey in the latter was, if anything, better than the former and the difference in price of the "get up" was only $1^1/_2$d.

10. Associations - To be successful combination is desirable and for beekeepers this want is met by the County Beekeepers' Associations. The help given to individual members is of various sorts, but it is worth the subscription to a beginner to have a visit from an expert. A man may read all there is to be read about bees and bee-keeping and then not be so well able to manage his bees as one who has seen the expert work amongst them. In the sale of our produce, too, the association often greatly helps. The hon. secretary of the Lincolnshire BKA sold over £50 worth of honey to one firm for the members last year. As this is one of the beekeepers' difficulties it should be sufficient to induce all beekeepers to join. In conclusion, if you want to be a successful beekeeper join your county association and loyally support its committee and officers.

BBJ 29 Mar 1900

The annual general meeting of members of BBKA was held on Thursday, the 22nd inst., in the Board-room of the RSPCA, Jermyn-street, SW. under the presidency of Mr. ED Till. Amongst those present was AG Pugh. A letter was received from Mr. P. Scattergood apologising for enforced absence.

Mr. Pugh thought the price of $4^1/_2$d quoted by Mr. Tolson was very low for section honey of good quality. As late honorary secretary of a county association, he had found the adoption of a county honey label of great service in promoting sales locally. In his opinion the question of the disposal of surplus was one to be dealt with by county associations rather than by the central body.

A general conversation ensued on the question of publishing and circulating leaflets as suggested by Mr. Spencer; the Chairman, Mr. Carr, Mr. Pugh, Mr. Brice, Mr. Taylor, and others taking part therein. In the end it was proposed, with apparently general assent, to endeavour to get the county associations to support the effort for issuing at a cheap rate a leaflet for distribution within their areas. An endeavour will therefore probably be made to carry it out in some form or other, have separate classes at shows for sainfoin and white clover honey respectively in sections. Messrs. Pugh, Carr and Hooker continued the discussion, which tended to confirm the Chairman's view.

NEP 5 Apr 1900

Welbeck Tenants Show. The arrangements for the Tenants' Show on June 5th, are almost finalised we have been informed by Mr. Hy. Woods (the secretary), Mansfield. As in the past two or three years, demonstrations of beekeeping will be given under the auspices of the NBKA.

BBJ 5 Apr 1900
Mr. Buller, Mr. Pugh, and Mr. Young deprecated the conveyance by rail of observatory hives and bees without a person in charge; the latter gentleman referring to an instance at the Royal Show where the railway company left such a hive exposed in the station yard for two hours in the sun, with the result that half the bees were found dead on arrival at the show ground.

Minute Book – Quarterly Committee Meeting held in the Peoples' Hall, Nottingham on April 8th, 1900. Present: Mr Ellis in the chair, Messrs Faulconbridge, Forbes, Harrison, Marriott, Pugh, Putterfield, Skelhorne, Swann, Wadsworth, Scattergood, Brooks, Hallam, Rawson, Hardy and Gray.

The minutes of the previous meeting having been read, the meeting confirmed them and they were signed off by the chairman. The quarterly balance sheet was next gone into and it was considered satisfactory. Mr Hardy proposed and Mr Wadsworth seconded the same be received and adopted.

It was resolved that the sub-committee for dealing with the grant from the County Council be composed of Messrs Ellis, Marriott, McKinnon, Puttergill, Pugh, Scattergood, Skelhorne and the secretary.

Mr Skelhorne proposed and Mr Scattergood seconded that we join the Moorgreen Show as usual and that Mr. Pugh be judge. It was also resolved that we join the Southwell Show on the usual terms and that Mr. Marriott be the judge.

As the Notts. Agricultural Society had again promised us the same terms as in previous years, it was resolved:
    that we hold the annual county show with them in Colwick on June 6th and 7th
    that the schedule shall be framed with 1889 classes and 1888 prizes as the basic or an increase in the prize list of 33/- over last year
    that Mr Cribb be asked to judge and lecture at the usual fee viz. 21/- per day and expenses or failing him Mr Howard
    that Mr Marriott be assistant judge

BBJ 11 Apr 1900
TML (Sheffield). Nottingham as a Bee District. Before moving to the lace county we recommend you to write to the Hon. Sec, who would no doubt advise you on the

several questions asked better than we can ourselves. Address: Mr. Geo. Hayes, Mona-street, Beeston.

BBJ 12 Apr 1900
For Sale, best offer, seven stocks of healthy bees in good frame-hives, four empty Frame Hives, Honey Extractor, Supers and worked-out frames of Comb, and lot of sundries; equal to new; leaving district. A. Temple, Aslockton, Notts.

BBJ 3 May 1900
Trade Catalogues Received.
EC Walton, Muskham — The new catalogue of bee-keeping appliances of the above old-established firm is but one of several comprehensive lists issued by them, the one before us being notable for the very moderate figure at which the goods are priced. This is, we understand, largely accounted for by the labour-saving appliances required for the extensive trade done in portable buildings of all kinds, varying in value from a poultry-house at 15s to building a greenhouse listed at nearly £300. They seem to do a very large and satisfactory trade, judged by the list of "testimonials," which latter alone fill over twenty pages.

BBJ 24 May 1900
Prepare for Your Honey Harvest — Wired Frames, "WBC" ends, fitted with "Weed" Foundation: shallow, 5s dozen; standard, 7s dozen. Will exchange for Stocks or Swarms. Massey, Farndon-road, Newark.

BBJ 31 May 1900
Natural Swarms for Sale, 10s each. Apply, H. Holleworth, Manor Farm, Wysall.

NG 9 Jun 1900
Notts. Agricultural Show. The exhibition promoted by the NBKA was distinctly successful and formed an attractive feature of the show. There was a capital entry for the prizes offered in the various classes and the specimens of honey exceeded expectations, the season being unusually late, and most of the honey of last year's production. Under the head of appliances, there was only one collection of hives and appliances and the judges did not feel justified in awarding the prizes. For the best and most complete frame hive for general use made by the exhibitor or his bona fide workmen, the first prize was secured by Mr. GH Varty, North Muskham, Newark, Messrs. Walton and Co., Newark, obtaining the second.

The first prize for the most attractive display of honey in any form and of any year, was awarded to Mr. GS Puttergill, who was also the recipient of the NBKA silver pendant, his specimens being admirably arranged. For light colour run or extracted honey produced in any year. Mr. AG Pugh (Beeston) was first, Mr J Herrod (Sutton-on-Trent) second, and Mr JT Faulconbridge (Bulwell) third. Mr. Pugh also received the BBKA silver medal and was awarded the first prize and BBKA bronze medal for dark colour run or extracted honey, the second going to Mr. P. Scattergood (Stapleford), and the third to Mr. Puttergill. Mr. Puttergill and Mr. Herrod took first and second respectively for sections of comb honey, Mr. Pugh being first in the class for granulated honey and taking the BBKA certificate. Mr Herrod being placed second and Mr. Geo Hayes (Beeston) third. For shallow frame of honey, suitable for extracting Mr W Swann (Eastwood) secured the first award and the NBKA silver pendant for an excellent frame, Mr. H. Merryweather (Southwell) coming second with only a slightly inferior exhibit for honey vinegar. Mr. Scattergood obtained first and the NBKA certificate, being similarly placed in the classes for honey cake and specimens of bees, Mr. Hayes running second in the two first-named classes and third for bees to Mr. Puttergill.

There were some nice specimens of beeswax, that staged by Mr. Puttergill being awarded premier place, whilst that of Mr. Scattergood was a good second. Amongst the exhibits not for competition, was noticeable a beautiful collection of lantern slides belonging to the NBKA and illustrating bee culture. The judging was kindly undertaken by Mr J Howard, of Holme, near Peterborough and Mr. SW Marriott (Sneinton). During the afternoon Mr. Howard gave demonstrations and lectures on the best methods of beekeeping and he was assisted by Mr. P. Scattergood.

BBJ 14 Jun 1900
XYZ (Newark-on-Trent). Foul brood is breaking out in comb sent.

BBJ 21 Jun 1900
Royal Agricultural Society. York Meeting, 1900.
Class 386. Honey Vinegar—1st, P. Scattergood, Stapleford, Notts.
Class 388. Interesting Exhibit of a Practical Nature—1st, P. Scattergood.

Minute Book – Quarterly Committee Meeting held in the Peoples' Hall on Saturday, July 7th, 1900. Present: Messrs Forbes Hallam, Marriott, Pugh, Puttergill, Rawson, Skelhorne, Swann, Turner, Wadsworth and secretary. Mr Rawson having been voted to the chair in the absence of Mr Ellis.

The minutes of the previous meeting were read, confirmed and signed by the chairman.

Various correspondence was next gone through about the arrangements for the annual show, etc.

The balance sheet for the quarter was next gone into and it was proposed by Mr

BEEKEEPING IN VICTORIAN NOTTINGHAMSHIRE

Pugh and seconded by Mr Wadsworth the same be passed. A discussion about awarding of BBKA medals and also of paying the 21/- for the same to the BBKA and whether it would not be advisable to purchase them ourselves and these matters were referred to the annual meeting.

The secretary next gave a report on the annual show at Colwick in which he stated that the entries got considerably less and it was felt that, unless members did more by way of exhibiting, it would not enable the committee to conscientiously ask for a renewal of the grant. He also stated that, at the show, Mr Puttergill reported losing from his trophy 15-20 lbs of honey. A discussion here followed and a feeling was expressed that this association should provide a watchman in future at this show and that the matter be referred to the annual meeting for consideration. The committee also decided (proposed by Mr Skelhorne and seconded by Mr Pugh) that 6/- be awarded to Mr Puttergill as part remuneration for his loss and also subscribed that amount from the meeting.

BBJ 12 Jul 1900
The annual exhibition of bees and honey was held in connection with the Leicestershire Agricultural Society's Show at Leicester on July 4 and 5. The honey tent was tastefully set out with plants and flowers kindly lent by Messrs. Boyes and Underwood. The unfavourable weather conditions during the past few weeks plainly told its tale and the show of honey was not up to the usual average. Lectures in the bee-tent were given at intervals by Mr. Riley, Leicester and Mr. AG Pugh, Notts. who officiated as judges.

BBJ 19 Jul 1900
The annual show of the Lincolnshire Agricultural Society was held at Spalding on July 12 and 13. Amongst prizes awarded was a collection of appliances - c, EC Walton, Muskham.

BBJ 2 Aug 1900
A meeting of the BBKA Council was held on July 26, at 12, Hanover-square, London, W. Mr WH Harris in the chair. A letter apologising for absence was received from P. Scattergood,

American Bee Journal, 2 August 1900
I thought it probable that the following may possess some interest for you or your readers.
A swarm of bees on the march.
When cycling this morning on my usual professional "round" I was not a little astonisht to see a swarm of bees walking in procession, like a long, brown snake, along the narrow footpath bordering the main road from here [Brant Broughton] to Newark.

The resemblance to what one could suppose Lord Robert's army on the march would appear like at once struck me. There were some few bees flying ahead, representing the "cavalry scouts;" then came the main army in serried ranks, extending to a length of several yards, all marching on foot – these were the "infantry," of course; and, finally, separated from the main body by about two feet, but with "scouts" passing to and fro, came a considerable cluster forming the indispensable "rear guard." A man working on the road informed me that the whole swarm had thus advanced about 20 yards since he had first observed them some time before.

I at once rode back to the house of a bee-keeper I knew who lived near, and failing to obtain a skep, got a box of shallow frames with comb built out and an old newspaper. Returning, I placed this "Pretoria" directly in front of the advancing army, covering the box with the newspaper and propping it up in front with a stone. I then continued my journey, and on my return found, as I had expected, that "the army" had "taken possession of the town," and that "all was quiet." This evening I drove over and took possession of the swarm, which I have now safely establisht in my apiary at home. Knowing, as we bee-keepers do, the loyalty of bees to their queen, it almost lookt as if these little wanderers had caught up the patriotic spirit of the day. Anyway, I have seen many swarms, but this is the first time I ever saw one *walk*.

Percy Sharp, Newark

(The above is not only interesting, but our correspondent's simile is a very happy one, there being little doubt that the queen's inability to fly kept the bees loyally marching on foot rather than take wing and desert her. Ed. British Bee Journal)
(*This article is reproduced exactly as written with the, to us, quirky spelling prevailing in the USA at the time. Is this our Newark – it seems so from the use of quotes from the Boer War and the fact that it first appeared in BBJ? Dr. Percy Sharp was from Lincolnshire and sometimes judged at our NBKA Honey Shows.*)

BBJ 2 Aug 1900
Dealing with Suspected Combs and Fertile Workers. New Zealand for Bee-Farming
1. I send two samples of comb marked "S" and "8" respectively, and will be obliged if you will say if either or both are foul broods. Sample marked "S" was taken a week ago from an empty hive in a neighbouring garden. The stock died in 1898, and the hive has been left *in situ* and open till I blocked up the entrance fourteen days ago, having been told that bees—probably my own—were busy in and about it. Samples marked "8" are from two combs from one of my own hives — one that has done no good this year (I bought it in April and found it queenless). It still remains queenless, despite introduction of frames with eggs and also a ripe queen-cell at intervals. The stock having lately become possessed of a fertile worker I wished to split it up amongst other stocks, as advised in the "Guide Book," but on examination fear it is diseased.

2. Can you give any information as to the pros, and cons, of New Zealand as a location for bee-farming,
   (a) Is it an industry of any importance there already?
   (b) If so, is it "domestic," so to speak, or organised in large concerns?
   (c) Are climatic conditions and native flora such as to make honey, &c., a fairly dependable crop?
   (d) Is any one district particularly suitable? HC Wallis, Old Colwick.

Reply.
1. Comb marked "S" from unoccupied hive is so old that all trace of brood has disappeared; the cell-cappings, however, clearly indicate foul brood. The other sample has no disease in cells, but it would be advisable to destroy the stock, seeing that the combs are occupied with drone-brood in worker cells and the bees, so long queenless, are so old and utterly worthless.
2. Some parts of New Zealand are good for bee-keeping, white clover being grown plentifully. We have no personal experience of NZ as a suitable place for bee-farming, but as large tracts of grazing land are sown with white clover good honey is produced there in some districts. The bee industry is not carried on by means of " large concerns," but much the same as in Canada; a few large apiaries are owned by farmers, while the majority of bee-keepers are (like our own people) in rather a small way. The honey crop there is, we believe, no more reliable than our own and, while some districts are, as we have said, good for honey, others are not suitable for bee-keeping by reason of plants and trees growing there which yield honey of poor quality.

Leicester Chronicle 11 Aug 1900
The Leicestershire BKA held their annual exhibition at the Abbey Park Flower Show on Tuesday and Wednesday. The show of bees and honey proved as interesting as in previous years. There was a large display of honey of a very superior quality in the exhibition tent. One class alone contained twenty exhibits, so many of them of such a high quality that the judges (Mr. HM Riley, of Leicester, and Mr. Hayes, of Beeston, Nottingham) had great difficulty in deciding their relative positions. But the great attraction was the tent in which the manipulation of bees took place. Lectures on bee management were also given on Tuesday by Mr. Hayes.

BBJ 16 Aug 1900
Bees and Bean-Blossoms.
Now that better forage is failing here bees are turning their attention a little to the scarlet-runner beans which are growing close to my hives. The several species of *Bombus* are the most attentive visitors and, either on account of their superior strength or longer tongues, they occasionally go to the front of the flowers to reach the honey, a method of procedure which our honey-bee seems to disregard when dealing with beans. It may be a matter of common knowledge amongst experienced

bee-folk, but until yesterday I was not aware that the blossoms of runner-bean are rifled in almost every case by means of a hole "chawed" through the under side of the calyx and that tubular part of the corolla which lies beneath it. I say "chawed" advisedly, as the hole is evidently not an organic contrivance of the flower, being of various shapes and sizes, with rough edges, which soon show signs of having been pinched and torn. Moreover, the two under petals, or "platform" of the flower, seem frequently to be withered before having served their evident function of levers for working the kind of piston-and-cylinder fertilising apparatus peculiar to these "butterfly" flowers. This withering is apparently due to the base of each petal having been eaten through which brings me at last to my questions as follows:

1. How is it that our beans are "setting" all right - enough of them at any rate - though the "chawed" calyx from which the little bean is generally hanging seems to show that some at any rate of our much belauded fertilising agents did not play fair? Are any other kinds of flowers mutilated in like manner?
2. If the percentage of unfertilised flowers is increased by this "chawing" process, who is responsible for the damage, *Bombus* or *Apis mellifica*? Perhaps neither, for I confess I have never caught either in the act.

HC Wallis, Old Colwick, Notts.

Reply.
1. The 'perforations' mentioned are found in all blossoms, notably in that of the field bean, of which the great majority of blooms will be found pierced at the base. It is generally understood that the *Bombus*, or humble bee, makes the perforations of which the hive bee avails itself in extracting the nectar otherwise out of reach.
2. In this case the *Bombus* decidedly.

BBJ 30 Aug 1900
Shropshire BKA Annual Show was held in conjunction with the Shrewsbury Horticultural Society's great floral fete on the 22nd and 23rd inst. in the beautiful grounds of 'The Quarry', Shrewsbury. There was a very good display of bee-produce, including splendid samples of extracted honey and some good sections.
Prize awarded for Honey Vinegar—1st, P. Scattergood

BBJ 6 Sep 1900
Artificial Manures - are they a Hindrance to Bees Working?
When judging at the Royal Counties Show at Windsor last year I was introduced to a gentleman in the tent where the exhibits were staged and from what I remember I believe that he was in some way or other connected with the firm of Messrs. Sutton, the well-known seedsmen of Reading.

Anyway, he was a botanist and a very interesting conversation we had. Among other things he asked if I "had noticed that bees will not work on plants or flowers if they have been stimulated with artificial manure?" I admitted that I had not, but he assured me that such was a fact and he had proved it many times. I made a note at the time, intending to test it for myself, but, as the matter is of considerable interest to "the craft" generally, I would like to know if any of your readers have noticed the same fact and, if they have, to give us the benefit of their observations in the Bee Journal?

Peter Scattergood, Stapleford, Notts.

Derby Daily Telegraph 13 Sep 1900
Derbyshire Agricultural and Horticultural Societies' 39th annual exhibition. The Derbyshire BKA held its 19th annual exhibition in connection to this event. Mr Peter Scattergood, Stapleford was the judge. He is a first-class expert of BBKA, and his decisions met with general approval.

BBJ 29 Sep 1900
The nineteenth annual exhibition of bees, hives, honey, &c., was held in connection with the Derbyshire Agricultural Society's Show at Derby on September 12 and 13. On Wednesday Mr. Scattergood, the judge, gave demonstrations in bee driving, etc., which were most interesting.
Prizes awarded for Beeswax. —2nd, H. Meakin, Newthorpe; 3rd, J. Stone.

NEP 30 Sep 1900
Agricultural Show at Moorgreen.
In somewhat oppressive weather the 49th exhibition was held on Tuesday. The bee and honey show was, as usual, under the auspices of the NBKA, which has several members in the locality. They demonstrated that the tuition they have received from the county association has been of service to them, their exhibits being excellent in quality and far ahead of those seen at many other shows. Had the season been a better one no doubt the quantity of honey shown would have been larger. It was felt that the county bee tent would have been an attraction and an effort is to be made to secure it for another year.

Minute Book – Quarterly Committee Meeting held in the Peoples' Hall on Saturday, October 6th, 1900. Present: Mr Puttergill in the chair, Messrs Trimmings, Faulconbridge, Harrison, Marriott, Pugh, Swann, Turner, Wadsworth, White, Brooks, Hallam and Gray.

The minutes of the previous meeting were read when Mr Pugh proposed and Mr Wadsworth seconded that the same be adopted. The quarterly balance sheet was gone through and Mr Turner proposed and Mr Swan seconded that the same be passed.
The secretary and others then gave a resumé of the various shows and after discussion Mr Gray moved and Mr Pugh seconded that we approach the Moorgreen societies with a view to the next annual show being held for one year at least.

Mr McKinnon's resignation which had been received by the secretary was read. When Mr Turner proposed and Mr Trimmings seconded that Mr McKinnon has taken this step and that the secretary write him saying they shall be sorry to lose him and ask him to reconsider before accepting his resignation.

BBJ 11 Oct 1900
The twenty-fifth annual show of the Dairy Farmers' Association opened at the Agricultural Hall, London, on the 9th inst., and continues till the 12th. The judges were Mr Walter F. Reid and Mr. Peter Scattergood.

BBJ 18 Oct 1900
AG Pugh and P Scattergood gave 2/6d each to the Basingstoke Bee Case

Testimonial to Mr. Hooker. We have pleasure is notifying the following addition to list of subscribers to the above fund published in our issues of October 17 and November 7 1900 respectively: P Scattergood

BBJ 18 Oct 1900
The monthly meeting of the BBKA Council was held on Thursday, the 11th inst. at RSPCA, Jermyn-street, W. In the absence, owing to indisposition, of Mr WH Harris (Vice-Chairman), Mr. EJ Till presided and there were also present AG Pugh, P. Scattergood.

The autumnal quarterly converazione was held as usual at the offices of the RSPCA. October 11th, when the spacious boardroom was crowded with bee-keepers attracted to London by the Dairy Show. Among the assemblage present were TS Elliot, AG Pugh and P. Scattergood.

Mr. Scattergood had very few remarks to make that were not commendatory. It was strange that so many persons sent dark honey to compete in the light honey class; sometimes it was vice versa. He wished the coloured glasses as arranged by the Secretary were well circulated, so that all members could know and adhere to the "standards."

There was a fine exhibit of queen-cells and queen-rearing which, unfortunately, was not staged until after the judges had left the hall, but it was a most interesting exhibit from a scientific standpoint. He was sorry they could not award it a prize. He had been looking through the list of exhibits and of counties from whence they came and noticed that many counties were not represented at all. He had talked this matter over with friends and from his own experience and information gathered he was led to the conclusion that beekeepers in the Midlands did not care to pay 2s 6d for the entry and 2s or 2s 6d more for the freight of an exhibit to the show, when they knew they had no chance of securing a prize. He advocated exhibits of a practical character, which brought before the public the various uses to which honey and wax might be put

The Chairman said Mr. Young, their Secretary, had taken great pains to make the question of colour clear to exhibitors. He had two pieces of different coloured glass prepared, which would indicate the proper colour for honey in the several classes. Mr. Young had taken care to send the glass samples especially to all exhibitors.

A short discussion then took place between Messrs. Brice, Scattergood, Reid and Meadows regarding the decisions of the judges at the show, which were to some extent challenged; but the questioner expressed himself amply satisfied with the explanations given.

Mr. Scattergood said that two years ago, at the corresponding conversazione, they missed the face of an old and valued friend, but sent him from that assembly a hearty fraternal greeting and he (the speaker) was sure he voiced the feelings of all now present when he said that they would be very glad to see Mr. Cowan back again as soon as he can come and that they hoped it would not be long before he returned to "dear old England." He (Mr. Scattergood) moved that that message be conveyed from the meeting to Mr. Cowan (loud applause). The motion was carried amid cheers.

After a few remarks from Messrs. Hooker, Scattergood, Ford and Reid on the subject of honey "sweating" through glass bottles, the Chairman expressed the acknowledgments of the meeting to the exhibitors of honey and appliances and a vote of thanks to himself for presiding concluded the proceedings.

Minute Book – Committee Meeting held in the Peoples' Hall, Heathcote-street, Nottingham on Saturday, January 5th, 1901. Present: Mr Ellis in the chair. Messrs Trimmings, Faulconbridge, Forbes, Harrison, Pugh, Puttergill, Skelhorne, Turner, Wadsworth, Scattergood, Brooks, W Herrod, Hardy and secretary.

The minutes of the previous meeting were read when Mr Scattergood proposed and Mr Skelhorne seconded the same be adopted.

A letter was received from Mr Brooks stating that the Moorgreen Societies would increase their grant from £3 to £5 and he also stated verbally that they would find the tent and staging for the annual show to be held there. Mr Ellis proposed and Mr Pugh seconded that we accept the offer and hold the annual show at Moorgreen.

Mr Pugh proposed and Mr Turner seconded that the prize list for the next annual show be the same as last year with the following alterations:

Class 3 Trophies – for the best and most attractive display of honey and wax in any form and of any year on space not exceeding 4ft wide by 2ft 6in deep and not to exceed 80 lbs of honey and wax suitable for a shop window display to be judged from the front only, weight to be stated. Flowers may be used for decoration.

A letter was read from Mr McKinnon stating that he had reconsidered his resignation and consented to remain a member.

The annual balance sheet was next gone through when Mr Scattergood stated that he had examined the accounts and found all quite accurate and proposed the same be received, adopted and printed on the notice of the annual meeting to the members. Mr Hardy seconded and it was carried unanimously.

The arrangements for the annual meeting were next gone into and it was arranged to hold the annual meeting at 3pm on Saturday, February 2nd in the Peoples' Hall, Heathcote-street, Nottingham and as the president had notified his inability to preside to ask the Mayor of Nottingham (JB Sim) to do so and failing him Mr Jesse Hind of Papplewick.
    that the tea be provided, as it was last year and that Mr W Herrod gives us about half an hour's paper after tea on the marketing of honey, etc.
    that there be one competition for a single jar of granulated honey – first prize to be 2/6d and the second prize 100 honey labels

The prize drawing was next proposed – Mr Ellis promising a hive if his Lordship's gift was not forthcoming Mr Trimmings a box of useful articles value 10/-, Mr Pugh one of Meadows Imperial rapid feeders and the association 200 honey labels.

Mr Pugh proposed and Mr Turner seconded that an invitation and ticket be sent to each of the three neighbouring association secretaries – Messrs Godson, Walker and Waterfield, for the meeting.

DEATH OF THE QUEEN

After a few days of anxious strain, felt not only by her own people but by the whole civilised world, our dearly loved and venerated QUEEN having passed away gently and peacefully surrounded by her children and grandchildren on the evening of Tuesday last.

BBJ Thursday 24th January 1901

www.ingramcontent.com/pod-product-compliance
Lightning Source LLC
Chambersburg PA
CBHW081203170426
43197CB00018B/2906